AF882109

Reprinted by St. John's Press
LLC
Alexandria, Va. 22301
ISBN 978-1-944961-22-0
2016

Introduction

The terrorist attacks on the World Trade Center and the Pentagon on 11 September 2001 dramatically changed the world in which we live. Never had an enemy attacked us on our own soil to such effect. Over three thousand Americans and several hundred foreign nationals from over ninety countries died that day at the hands of a ruthless, and to some degree faceless, enemy. The terrorist organization known as al-Qaeda, perpetrator of the attack, operated in the shadows to take advantage of the freedom and openness that are American hallmarks. Afghanistan, a known training ground and a safe haven for al-Qaeda, quickly became the focus of the first military efforts to strike back. Osama bin Laden, al-Qaeda's enigmatic leader, believed he and his followers were beyond the reach of American arms in that far-off mountainous land, protected by its fanatical Taliban regime.

Bin Laden was wrong, and America reached deep into Central Asia to find his organization and neutralize it. Most Americans are familiar with the military operation that subsequently took place in Afghanistan. In a matter of months, the U.S. Army, Air Force, Marine Corps, and Navy, in a masterful display of joint operations and in concert with our Afghan allies, overthrew the Taliban regime and drove the terrorist al-Qaeda into worldwide flight.

This pamphlet provides details on the role of the U.S. Army in the critical three-year period following the conclusion of Operation ANACONDA in March 2002. It details the story of American and international forces working to solidify the initial invasion's crippling of al-Qaeda and removal of the Taliban. It recounts the story of the quest to build a new, democratic Afghan government capable of maintaining internal security and tending to the needs of the Afghan people. It tells the tale of the U.S. Army's search for a proper balance between counterterrorism and counterinsurgency operations as the enemy rebuilt his forces from safe havens in Pakistan. Finally, it chronicles the Army's efforts to maintain an effective presence in Afghanistan while juggling the challenges of an indigenous population historically opposed to foreign forces and the decreased resources available after the start of the Iraq war in 2003.

Operation ENDURING FREEDOM, March 2002–April 2005, was written by historians Dr. Brian F. Neumann, Dr. Lisa Mundey, and Dr. Jon Mikolashek in the Histories Division of the Center of Military History. We hope that you enjoy and profit from this dramatic but often overlooked story of our Army in action in the Global War on Terrorism. It is the second in the Center's series of campaign brochures on the U.S. Army in Afghanistan.

RICHARD W. STEWART
Chief Historian

The U.S. Army in Afghanistan Operation ENDURING FREEDOM March 2002–April 2005

In a fast-paced, low-cost campaign in Afghanistan from October through December 2001, small numbers of U.S. troops backed by U.S. airpower joined local militias to topple the Taliban regime and damage the al-Qaeda network. Thus began Operation ENDURING FREEDOM (OEF) and the fulfillment of President George W. Bush's promise after the 11 September attacks on the United States to wage war on the perpetrators and their protectors. For the next three years, while also striking in Iraq, the United States sought to create a viable Afghan government and complete the dismantling of the Taliban and al-Qaeda. It would take the full three years, however, for the Americans and their allies to realize just how complex the issues in Afghanistan were and to devise a plan to address them.

Strategic Setting

With the collapse of the Taliban regime in late 2001 and subsequent operations by American, British, Canadian, Australian, and Afghan forces in the first three months of 2002, ENDURING FREEDOM appeared to be on track. In December 2001, the United Nations sponsored a meeting of prominent Afghans in Bonn, Germany, to establish a plan for constituting a new national government. The group agreed to establish a six-month Afghan Interim Authority, to be followed by a two-year Afghan Transitional Administration. During the latter period, a *loya jirga* (a gathering of tribal dignitaries to select national leaders or approve government changes) would meet to adopt a constitution and arrange for elections. The Bonn meeting selected Hamid Karzai—scion of a prominent Pashtun family, a former deputy foreign minister, and leader of a small band of militia fighters during the overthrow of the Taliban—to serve as chairman of the interim authority. While the Bush administration supported this process, it sought to do so without becoming enmeshed in a drawn-out military intervention or engaging in nation building.

Yet, for all the administration's hopes, the situation in Afghanistan was far from stable. Decades of civil war, religious division, and economic malaise had left the country the very embodiment of a failed state. Ethnic differences in the population, consisting of

Pashtuns (42 percent), Tajiks (27 percent), Hazaras and Uzbeks (9 percent each), and various smaller groups, bred competition and discord. Provincial governors, some of whom were warlords rewarded for their part in toppling the Taliban, operated semi-independently, if they were at all able to establish control in their regions. At times, President Karzai seemed to hold sway only in the capital of Kabul. Indigenous security forces capable of providing stability did not exist, nor did a sense of national unity or identity that could bind the competing groups together.

The Taliban and al-Qaeda, though forced underground or across the border into Pakistan after their defeats at Tora Bora in December 2001 and in Operation Anaconda in March 2002, also remained active. Al-Qaeda took root primarily in North and South Waziristan in Pakistan's semi-autonomous Federally Administered Tribal Areas. From there, fighters crossed the border at will into the Afghan provinces of Nuristan, Kunar, Nangarhar, Paktiya, Paktika, and Khost. The Taliban sought refuge in and around the city of Quetta, in Pakistan's Balochistan Province, where it began to rebuild. Rank-and-file members of the Taliban who could not escape to Balochistan melded into the Afghan population, generally in Uruzgan and Kandahar Provinces where their support among the Pashtuns remained strong. There they waited, determined to undermine efforts to build support for the emerging central government until such time that their movement could go on the offensive.

During the late spring and early summer of 2002, other groups joined the Taliban and al-Qaeda for either nationalistic or religious reasons to oppose the presence of foreign troops in Afghanistan. The most prominent of them proved to be Hizb-e-Islami Gulbuddin, led by Gulbuddin Hekmatyar. A militia leader during the war with the Soviets in the 1980s, he served as prime minister in one of Afghanistan's warlord-based governments prior to the Taliban's rise. Operating in northeastern Afghanistan, Hekmatyar's fighters seldom engaged in open combat with the Americans and their allies, favoring indirect means of laying mines or launching rockets. Narcotics traffickers also resisted incursions into poppy-growing areas.

Geopolitics fed Afghanistan's instability. For centuries the region had been a crossroads of cultures and an avenue of invasion. More recently, Iran supported the Hazaras, a Shi'ite group. Saudi Arabia provided aid to the Sunni Muslims of the Taliban. And China, Russia, and India all had economic and political interests in the war-torn country. (*See Map 1.*)

Pakistan's long-term conflict with India made it acutely concerned with Afghanistan. Pakistan and India fought four wars between 1947 and 1999, mostly over the disputed Kashmir region, and almost fought a fifth in 2002. Fearing strategic encirclement, the Pakistanis preferred instability on their western border over a coherent Afghanistan in India's camp. Pakistan supported the mujahideen following the 1979 Soviet invasion of Afghanistan, where Pakistan's Inter-Services Intelligence Directorate, aided by the U.S. Central Intelligence Agency (CIA), provided assistance to Afghan leaders like Hekmatyar and Jalaluddin Haqqani. After the Soviet withdrawal, Pakistan continued to furnish support to Muslim fundamentalist groups in the region, including the Taliban after its emergence in the early 1990s.

Ethnicity and religion complicated relations between Afghanistan and Pakistan. Pashtuns, the largest ethnic group in Afghanistan, were the second largest ethnic group in Pakistan, making up just over 15 percent of the population. They constituted a majority in Pakistan's western territories and provinces that border Afghanistan. Religious schools, or *madrassas*, located in northwestern Pakistan offered free education to hundreds of thousands of Afghan and Pakistani Pashtuns. The Taliban grew out of these schools, many of which taught a radical version of Sunni Islam. Pakistan and Saudi Arabia provided vital assistance to the schools and expanded that support to include the Taliban. The Pakistani government used the fundamentalists as a source of opposition to expanding Indian, Iranian, Russian, and Chinese influence in the region.

The Pakistanis did not believe their relationship with the Taliban was incompatible with their alliance with the United States, which they saw as a useful strategic counterweight to China's links with India. Pakistan allowed the U.S. military to use its territory as a logistical base in exchange for financial aid and access to modern weaponry. Yet after the Taliban government's collapse, Pakistan continued to support Muslim fundamentalist groups, including Hizb-e-Islami Gulbuddin and the Haqqani network. Indeed, factions within the Pakistani government often worked at cross-purposes, with one group supporting the Americans while another offered aid to insurgent groups operating within Pakistan. The schizophrenic nature of Pakistan's policy settled into a pattern where Pakistan would go aggressively after al-Qaeda but only selectively target the Taliban in order to preserve it as an option for future influence in Afghanistan. The quandary for the United States was obvious. Pakistan provided America's lifeline in Afghanistan and sanctuary for America's enemies.

Karshi-Khanab
UZBEK
TURKMENISTAN
IRAN
JOWZJAN
Mazār-e
BALKH
Sar-e Pul
Maīmanah
FARYAB
SAR-E PUL
BADGHIS
BAMYA
GHOR
Herāt
HERAT
Kakarak
DAYKUNDI
Shīndand
URUZGAN
FARAH
Tarīn Kōt
ZABUL
HELMAND
Lashkar Gāh
Kandahar Airfield
IRAN
Spīn Bōldak
NIMROZ
KANDAHAR

Map 1

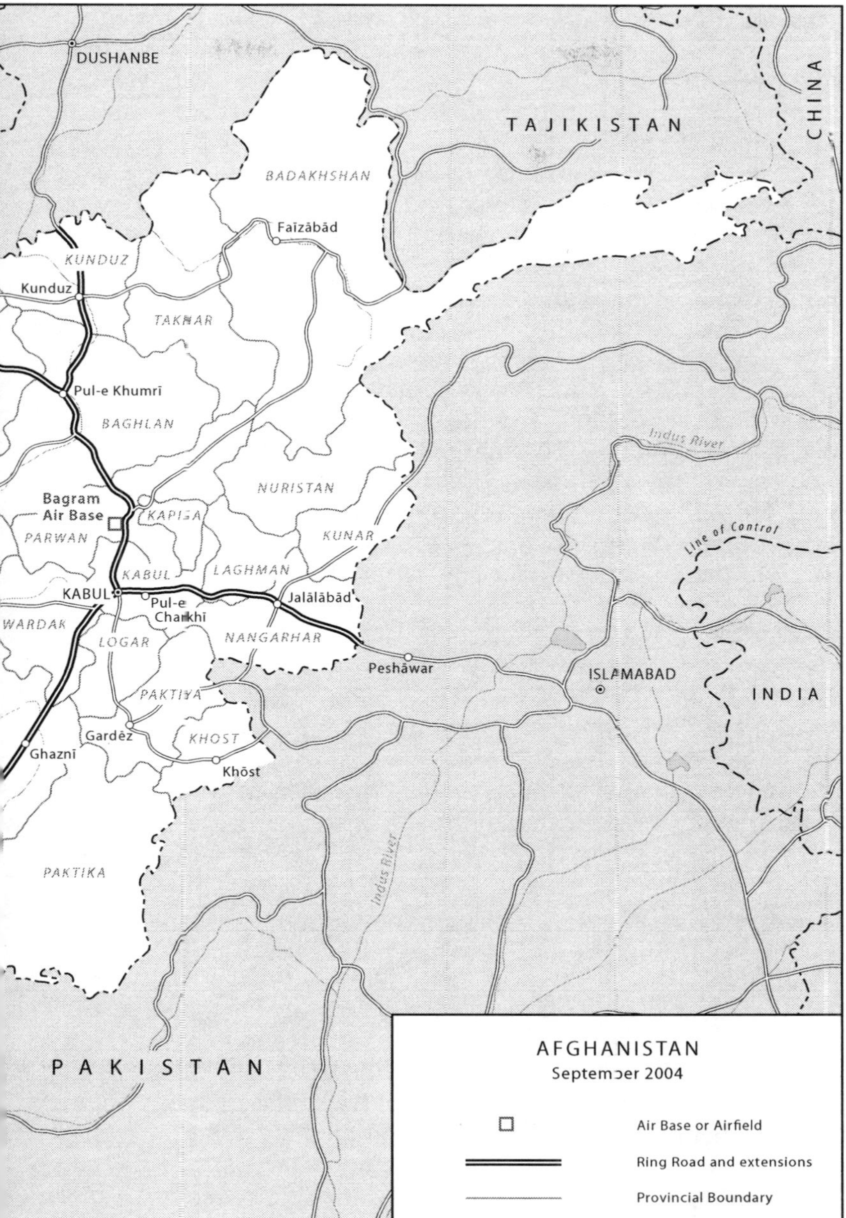
DUSHANBE
TAJIKISTAN
CHINA
BADAKHSHAN
Faizābād
KUNDUZ
Kunduz
TAKHAR
Pul-e Khumrī
BAGHLAN
Indus River
NURISTAN
Bagram
Air Base
KAPISA
PARWAN
KUNAR
Line of Control
KABUL
LAGHMAN
KABUL
Pul-e
Charkhī
Jalālābād
WARDAK
LOGAR
NANGARHAR
Peshāwar
ISLAMABAD
INDIA
PAKTIYA
Gardēz
KHOST
Ghaznī
Khōst
PAKTIKA
Indus River
PAKISTAN
AFGHANISTAN
September 2004
Air Base or Airfield
Ring Road and extensions
Provincial Boundary
0
150
Miles
0
150
Kilometers

Consequently, the United States faced a situation that defied easy solution. The previous six months had witnessed military and diplomatic progress. By March 2002, Operation Enduring Freedom had produced a coalition of fifty-six nations that furnished support ranging from the commitment of combat troops to logistical aid. American, British, Canadian, and Australian troops formed the major share of coalition fighting forces, with the largest contingent from the United States. In addition, in December 2001 the United Nations authorized a small security element under British command, the International Security Assistance Force (ISAF), to protect Kabul and its environs. The United States did not participate in the effort, and ISAF was not initially a part of the coalition. The total foreign troop commitment numbered some twelve thousand, of which the United States contributed about half. But there were no plans to increase that contingent. As Secretary of Defense Donald H. Rumsfeld later explained, "Long-term stability comes not from the presence of foreign forces but from the development of functioning local institutions." The United States would aid Afghans in creating those institutions without, it hoped, making large-scale deployments to the country.

Operations

Lt. Gen. Paul T. Mikolashek's Third Army headquarters, serving in Kuwait as the Combined Forces Land Component Command (CFLCC), assumed responsibility for Afghanistan in late 2001. Special Operations Forces supplied much of the American presence on the ground during the war's opening months, but, as that fight expanded, more conventional units arrived in Afghanistan. The Afghan landscape played a critical role in defining what type of forces the United States deployed. Roughly the size of Texas, Afghanistan can be divided into five regions. The Hindu Kush Range begins in the northwest and runs to the southwest, creating a large plateau that bisects the country. The capital city of Kabul rests on the southeastern edge of this plateau. The eastern and northern sections of the country are mountainous and defined by long, thin valleys and relatively arid terrain. The western regions are more fertile lowlands bordering Iran. The southern provinces running between the Hindu Kush and the Pakistan border are characterized as scrubland or desert. The country's size and dispersed population required a mobile force to provide effective security. The difficult terrain meant helicopters were necessary to reach the isolated rural population, particularly in the mountainous regions along the Pakistan border in the east. Therefore, the United States needed a light, mobile force that could conduct operations and be inserted, supplied, supported, and evacuated by helicopter.

Maj. Gen. Franklin L. Hagenbeck's 10th Mountain Division represented just the type of light force that the situation in Afghanistan demanded. Its headquarters deployed in December from Fort Drum, New York, to Karshi-Khanabad Air Base in Uzbekistan to take control of ongoing operations. Dubbed CFLCC-Forward or Task Force Mountain, Hagenbeck's headquarters moved in February 2002 to Bagram Air Base, a former Soviet base thirty-five miles north of Kabul, to oversee planning for Operation Anaconda, a large-scale mission to clear out the remaining Taliban and al-Qaeda forces from Paktiya Province in eastern Afghanistan (*Chart 1*).

By the time Task Force Mountain reached Bagram, U.S. forces in Afghanistan had grown significantly. The 10th Mountain Division's main contribution consisted of Task Force Commando, formed around the 2d Brigade headquarters, with one infantry battalion, three aviation battalions, and attached special operations units. It joined Joint Special Operations Task Force–North, composed of the 5th Special Forces Group; the 19th Special Forces Group; and the 2d Battalion, 160th Special Operations Aviation Regiment. Joint Special Operations Task Force–South, which included the 3d Special Forces Group and the 3d Battalion, 160th Special Operations Aviation Regiment, arrived in January 2002. Task Force Rakkasan, which was built around the 3d Brigade, 101st Airborne Division, and contained three infantry battalions, three aviation battalions, and the 626th Support Battalion, landed in theater at the same time. In March, U.S. Central Command (CENTCOM) concentrated command authority over special operations units in Afghanistan by creating the Combined Joint Special Operations Task Force (CJSOTF) and placed it under Task Force Mountain's tactical control.

With the completion of Operation Anaconda, the U.S. military sought to consolidate its successes and to help the Afghans create civil and military institutions without increasing its military presence in Afghanistan. Additional coalition forces arrived as allied nations contributed troops to the mission of rebuilding the war-torn country. In April 2002, the United Kingdom added the 45 Commando, a battalion-size unit of about six hundred Royal Marines. Canada deployed the 3d Battalion, Princess Patricia's Canadian Light Infantry, as well as special forces soldiers. Romania sent the 26th Infantry Battalion, which provided base security for Kandahar airfield. Several coalition partners deployed engineer units, including Poland,

Chart 1—Combined Forces Land Component Command–Forward Rotation, February–May 2002

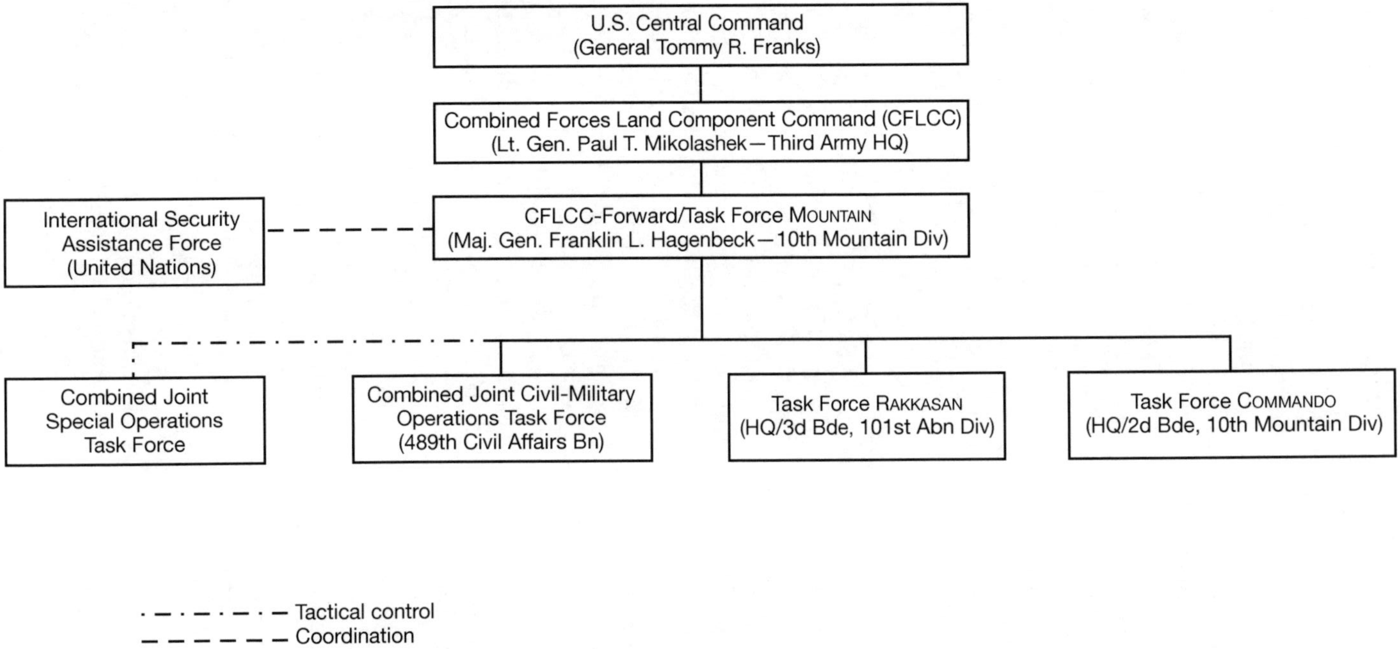

Norway, Slovakia, and Italy. The Poles and Norwegians cleared mines in Bagram, while the Slovakians and Italians worked on the airfield and runway. Meanwhile, U.S. Army logisticians established more permanent ground lines of communications. The coalition's overland supply system into Afghanistan ran through Karshi-Khanabad in the north and the port of Karachi, Pakistan, in the south. As the number of coalition forces in Afghanistan grew, these supply routes became even more important.

Though the United States did not commit itself to long-term engagement in Afghanistan, it did decide to continue operations after ANACONDA, which began to strain Task Force MOUNTAIN's capabilities. Essentially a stripped-down division headquarters, the task force proved increasingly ill-suited to coordinate all conventional and unconventional operations taking place across Afghanistan. As the number of American forces jumped to over five thousand by March 2002, the task force headquarters grew more overstretched. Its geographic responsibility expanded to encompass not only Afghanistan, but also Tajikistan, some of Uzbekistan, and parts of Pakistan. U.S. commanders understood that they were operating under an informal force-cap, established by CENTCOM and the Defense Department, of around seven thousand personnel. Such strictures required flexibility not only of combat units, which would be called on to perform a wide variety of missions, but also the command and control apparatus required to oversee those operations. The challenges facing MOUNTAIN increased when CFLCC headquarters in Kuwait began devoting more effort to planning a potential campaign against Iraq. With authority for operations and responsibility for missions divided between CENTCOM, CFLCC, and Task Force MOUNTAIN, it soon became clear that the Army required a more simplified and capable system to stabilize the situation in Afghanistan.

Operation MOUNTAIN LION

Despite Operation ANACONDA's success in dislodging the Taliban and al-Qaeda from the Shahi Kot valley in Paktiya Province, U.S. and coalition forces needed to maintain the pressure in southeastern Afghanistan to prevent the enemy from interfering with the upcoming loya jirga intended to create a new transitional government. Planning for what would be Operation MOUNTAIN LION began in March 2002. Envisioned as a three-month operation composed of week-long missions launched by helicopter from Bagram and Kandahar, MOUNTAIN LION sought to identify, isolate, and destroy al-Qaeda and Taliban forces as well as to deny them sanctuary and freedom of movement in Paktiya, Paktika, and Uruzgan Provinces. The operation featured a new set of tactics that relied less on Afghan allies and airpower and more on well-trained coalition light infantry. Begun in April and running through June, the operation's key tasks included isolating important border areas, locating the enemy, conducting reconnaissance and raids, and patrolling the border with Afghan militia forces. In order to undermine al-Qaeda's and the Taliban's influence on the Afghan people, General Hagenbeck directed coalition forces to provide humanitarian assistance. He also encouraged the work of private organizations, such as the Red Cross and Care International, and the Afghan government to improve living conditions for the population.

MOUNTAIN LION involved several thousand American, British, and Canadian troops from conventional units, as well as American and Australian special operations forces, operating in southeastern Afghanistan. MOUNTAIN LION kicked off with U.S. troops from the 101st Airborne Division, along with a battalion-size task force of Royal Marines, being airlifted into Paktika Province. While conventional forces failed to make contact with the enemy, on 30 April a reconnaissance patrol from the 2d Australian Special Air Service Regiment encountered several Taliban near Khost. After killing two enemy fighters, the Australians were reinforced by a rifle company of the 101st Airborne Division dispatched from Bagram. Although the appearance of coalition reinforcements drove the Taliban survivors into hiding, U.S. soldiers searched the surrounding area and uncovered several caches of weapons and ammunition.

Canadian troops joined the operation in early May, participating in an air assault on Tora Bora in Nangarhar Province to gather intelligence from al-Qaeda cave complexes abandoned months earlier. In addition to the operations conducted by conventional units, U.S. Special Operations Forces established small forward bases in three eastern Afghan provinces. The teams recruited local militias to disrupt Hizb-e-Islami Gulbuddin fighters operating in Kunar Province. In Nangarhar Province, they coordinated with the local warlords to target terrorists seeking sanctuary in Jalalabad, a city of about two hundred thousand on the main road from Kabul to Peshawar, Pakistan. In Uruzgan Province, Special Forces teams and Afghan militia conducted patrols to keep the enemy from reestablishing a foothold in the region.

Combined Joint Task Force–180 Takes Over

In April 2002, General Tommy R. Franks, head of CENTCOM, addressed the issue of command and control in Afghanistan by deciding to deploy a corps-level headquarters to Bagram. Franks assigned the mission to Lt. Gen. Dan K. McNeill's XVIII Airborne Corps, redesignated as Combined Joint Task Force–180 (CJTF-180). While primarily an Army organization, it included members of the U.S. Air Force and Marine Corps, as well as British, Australian, Polish, South Korean, Romanian, and other coalition personnel, making it a joint and combined staff. Although a typical combined joint task force could consist of a thousand personnel, CJTF-180 arrived in May 2002 at Bagram from Fort Bragg, North Carolina, with only 365 people in order to comply with personnel restrictions on U.S. forces in Afghanistan. The transfer of authority from Task Force MOUNTAIN to CJTF-180 occurred on 1 June 2002. Task Force MOUNTAIN did not disappear but instead assumed a subordinate role nested within McNeill's headquarters.

CJTF-180 had three other subordinate commands. A medical task force provided assistance to coalition forces and helped build Afghan medical programs. The Aviation Task Force (composed of Army and Marine air units) supplied helicopter and fixed-wing support, projection of combat power, and a quick reaction capability throughout Afghanistan. Finally, the Joint Logistics Command managed support within the area of operations. CJTF-180 also exercised tactical control of CJSOTF, though Central Command's special operations component retained its administrative and logistical responsibilities for Special Forces units operating in Afghanistan. Composed primarily of U.S. Special Operations Forces, CJSOTF included special operations elements from Australia, Canada, Denmark, France, Germany, New Zealand, Norway, Poland, the United Kingdom, and others at one time or another between 2001 and 2005.

While the command structure evolved, the mission for U.S. forces in Afghanistan remained the same. CENTCOM ordered CJTF-180 to eliminate the remaining al-Qaeda and Taliban forces in the country, to train the Afghan National Army, and to conduct operations to stabilize and secure Afghanistan so that terrorist organizations could not reemerge there. Central Command continued to operate under the assumption that McNeill's headquarters would turn over responsibility for security, civil-military operations, and humanitarian assistance to the Afghans within twelve to eighteen months, though no specific timeline existed (*Map 2*).

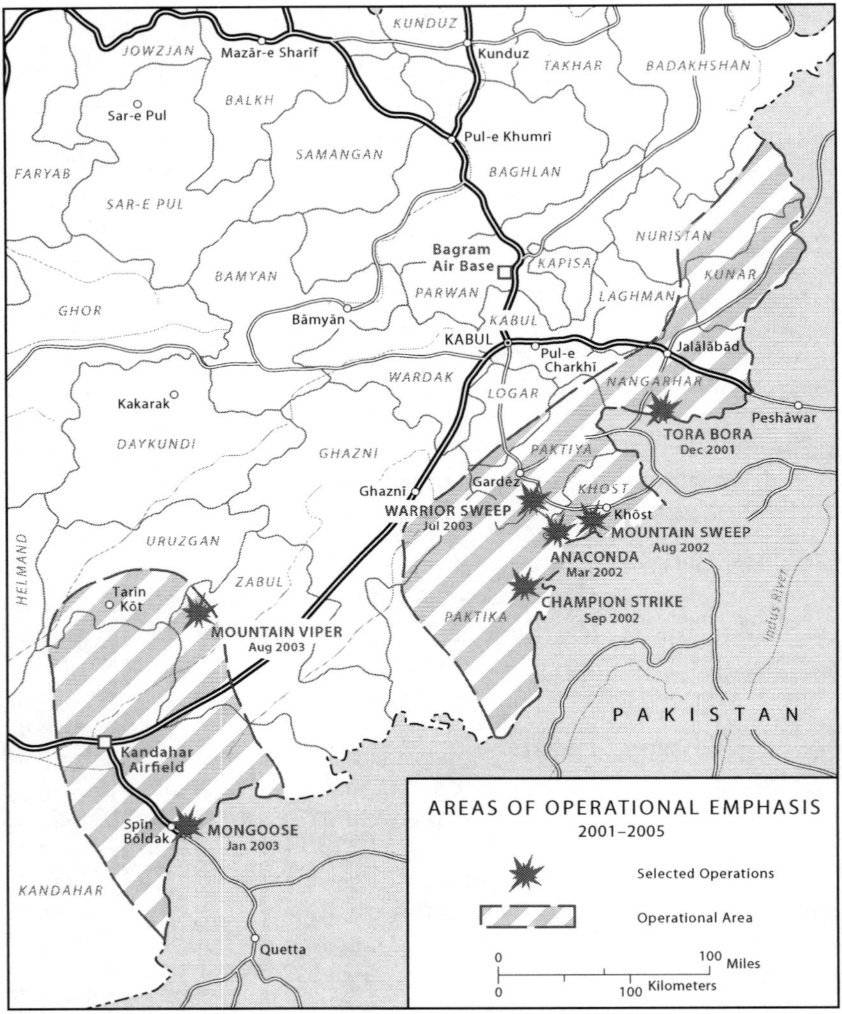

Map 2

The operational realities that emerged during the spring of 2002 made eliminating al-Qaeda and the Taliban difficult for CJTF-180. To do so, coalition forces needed to cross into Pakistan, which international law and the rules of engagement prohibited. Instead, McNeill focused on denying sanctuary and restricting freedom of movement within Afghanistan for the remnants of al-Qaeda and the Taliban. In particular,

U.S. forces hoped to prevent the enemy from using trails across the Afghan-Pakistan border, dubbed rat-lines, which smugglers often traveled. The difficult terrain meant that the paths were passable only by foot or pack animal. U.S. forces also sought out caches of weapons and ammunition, taking these munitions from the enemy and giving them to the Afghan National Army.

The inability of U.S. forces to pursue the enemy into Pakistan did not mean that the region offered complete safety to al-Qaeda. After joining the American coalition, and especially in the wake of Tora Bora, the Pakistani military and intelligence services increased their efforts to locate and arrest members of al-Qaeda. With the aid of U.S. Special Operations Forces and law enforcement agencies, including the CIA, the Federal Bureau of Investigation (FBI), and Pakistan's Inter-Services Intelligence Directorate, the Pakistanis arrested dozens of fugitive fighters, most notably Abu Zabaydah. Despite these successes, American authorities had their doubts regarding the Pakistani government's commitment to defeating the Afghan Taliban. On a positive note, MOUNTAIN LION prevented insurgents from disrupting a meeting of Afghan tribal leaders to form a new government to replace the interim authority. On 19 June, the United Nations announced the creation of the Afghan Transitional Administration (also known as the Afghan Transnational Authority), with Hamid Karzai as president.

Though coalition operations succeeded in keeping pressure on the enemy, they also had unintended consequences. On the night of 30 June 2002, a Special Forces team conducting reconnaissance near the village of Kakarak reported antiaircraft fire directed at nearby U.S. aircraft. The soldiers on the ground called in the coordinates of suspected enemy positions to an AC–130 Spectre gunship orbiting overhead. A B–52 Stratofortress also responded, dropping a Joint Direct Attack Munition precision bomb. The next day, the villagers asserted the air strikes hit a wedding party, killing forty. The provincial governor claimed that the Americans mistook celebratory gunshots for antiaircraft fire or that a rival tribe misinformed the Special Forces team. However, Army civil affairs personnel responding to the incident found antiaircraft artillery in the vicinity. They treated some injured Afghans but could not identify anyone killed by air attacks. Nevertheless, the incident resulted in unfavorable Afghan and international opinions concerning U.S. operations in Afghanistan. News stories focused on the reported civilian deaths and not on the U.S. troops' belief that they had been under attack. The experience highlighted the importance to the U.S. military of supplying timely and accurate facts. If it did not, as the bombing story indicated, reporters would seek out stories from other sources.

While the battle for U.S. and global opinion mattered, the more important issue was the reaction of Afghans to these events. American soldiers believed they were justified in using airpower to attack legitimate targets in a way that limited their own casualties, even if the enemy intermingled with civilians. The local population, however, remained disinterested in the finer points of U.S. fire support doctrine. The people objected to the loss of lives and property and the general behavior of coalition forces. House-to-house sweeps, the detention of men suspected of supporting the Taliban or al-Qaeda, and searches of women all added to growing tensions. Making matters worse, such actions sometimes were the result of baseless allegations made by one tribe or faction against rivals. These activities tarnished the image of the U.S. military in the country and CJTF-180's relations with the population.

Combined Joint Task Force–82 Arrives

In the summer of 2002, elements of the 82d Airborne Division under Maj. Gen. John R. Vines began rotating into Afghanistan to replace Task Force Mountain. Designated Combined Joint Task Force–82 (CJTF-82), it assumed responsibility for tactical operations. Two infantry battalions from Col. James L. Huggins' 3d Brigade, along with a battalion from the 1st Brigade, became the main maneuver force, designated Task Force Panther. The arrival of a sizable element of the 82d Airborne Division increased the overall number of U.S. forces in the Afghan theater from seven thousand in May 2002 to over nine thousand by the end of August (*Chart 2*).

While military units rotated in and out of Afghanistan, civilian contractors remained and took on an increasing amount of work. In 2003, when a forward support battalion from 10th Mountain Division arrived at Kandahar airfield, Kellogg, Brown, and Root—a Halliburton subsidiary—already ran the laundry and performed camp maintenance. Contractors expanded into food preparation, as well as staffing the warehouse, supply point, and other facilities. Civilian engineers, carpenters, masons, electricians, and plumbers also supported base operations. Despite their increased workload, contractors did not venture out into the country, leaving the most dangerous tasks to soldiers.

For combat units in Afghanistan from mid-2002 until autumn 2003, their basic mission statement remained to deny sanctuary to the enemy, disrupt the ability of al-Qaeda and the Taliban to plan

Chart 2—Combined Joint Task Force–180: First Rotation, June 2002–May 2003

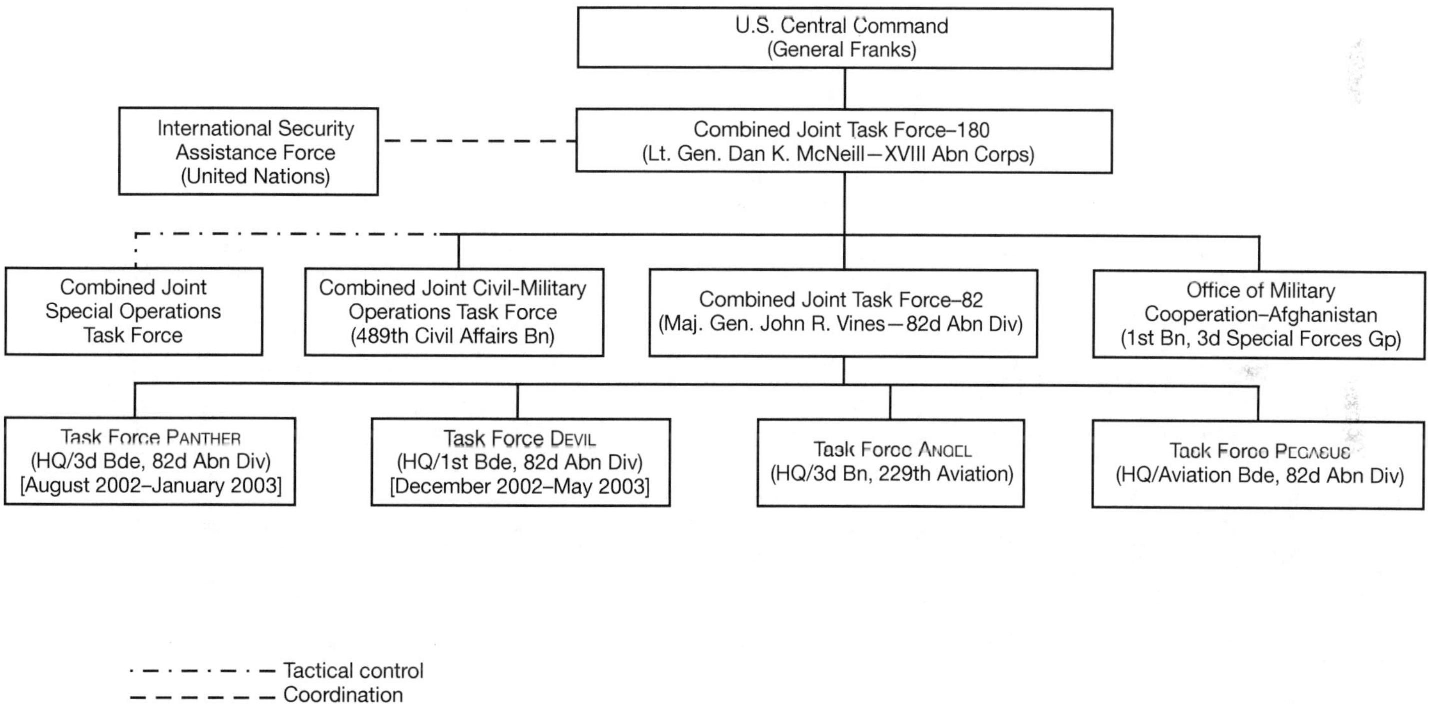

and execute operations, and destroy enemy forces when in contact. Coalition operations followed a typical template of search and attack within a defined area. Inside a designated box, troops methodically searched villages after cordoning them off. On occasion these were battalion-size missions, though more often smaller forces such as companies or platoons were used. In addition, twelve-man Special Forces teams worked with companies of Afghan troops and accompanied coalition battalions on the cordon and search missions.

Bagram Air Base and Kandahar airfield served as the primary bases for these units with smaller elements spread out in forward operating bases. When given a mission, the troops typically conducted an air assault by helicopter into an area before proceeding on foot for the next two or three weeks. They sent any detainees back to Kandahar or Bagram for interrogation. At the conclusion of the operation, the unit pulled back to its base, debriefed, refitted for a week to ten days, and then went out on the next mission, which could be in an entirely different area.

The Taliban responded to these cordon and search missions by attacking isolated operating bases with rocket-propelled grenades or mortar rounds rather than direct assaults on coalition units in the field. The outposts at Lwara and Shkin in Paktiya and Paktika Provinces, respectively, endured several attacks in the latter half of 2002 because they were close to the border with Pakistan. Although helicopters attempted to block the enemy retreat into the Pakistani sanctuaries, the difficulties associated with spotting individuals moving under tree cover in rugged terrain usually allowed the Taliban fighters to escape with impunity.

A few notable missions exemplified the war during this period. In August 2002, CJTF-82 launched MOUNTAIN SWEEP, which targeted al-Qaeda and Taliban forces in Paktiya and Khost Provinces on the Pakistan border. A task force of two thousand soldiers, formed around Lt. Col. Martin Schweitzer's 3d Battalion, 505th Infantry; several companies of Army Rangers; combat engineers; civil affairs teams; coalition special operations forces; and Afghan militia forces, took part in the operation. The maneuver began with the construction of a forward operating base near the border by personnel from the Louisiana National Guard and the 307th Engineer Battalion (Airborne). The week-long mission began on 18 August with an air assault by the 3d Battalion, 505th Infantry, using CH–47 Chinook helicopters from Kandahar airfield. After their insertion, Schweitzer's troops moved from village to village across the Zurmat District, southwest of the town of Gardez. The Americans discovered anti-

Inserting troops via Chinook helicopter, March 2002

aircraft artillery, weapons, and ammunition and captured ten men with suspected al-Qaeda or Taliban ties.

While CJTF-180 called MOUNTAIN SWEEP a success, Special Forces detachments in Khost Province faced a much more hostile population after the campaign. The special operations soldiers believed that the tactics used by conventional units undermined the rapport they cultivated among the local populace. The mission also failed to achieve the capture of a suspected al-Qaeda financier, who fled to Pakistan just prior to the action's commencement. "I have no doubt that they had advance warning that we were coming," Colonel Huggins stated. "We have to share information with the country we're in. I have no idea what they do with that information." Though Huggins saw no deliberate activity by Afghans that undermined the operation, his concerns reflected the continuing difficulties in coordination, communication, and intelligence sharing between coalition and Afghan forces.

In September 2002, the 82d Airborne Division launched another big operation, CHAMPION STRIKE. Approximately one thousand troops, including Schweitzer's 3d Battalion; elements of the newly arrived 1st Battalion, 504th Infantry, led by Lt. Col. David T. Gerard; Special Forces teams; and Afghan militia, conducted air assaults into the Bermail valley of Paktika Province. They captured

A female American soldier searching two Afghan women

an al-Qaeda or Taliban financier along with other suspects and also uncovered weapons caches and Taliban documents. Female military police soldiers discovered that some Afghan females were concealing weapons and ammunition under their full-length garments (called burkas), probably because enemy fighters thought they would not be searched. Changes in U.S. troop behavior in these searches did not produce the same disruptions as in previous operations.

At the end of 2002 as winter weather set in, CJTF-180 and CJTF-82 prepared to rotate their maneuver forces. Task Force PANTHER gave way to Col. John F. Campbell's Task Force DEVIL. Formed around the 1st Brigade, 82d Airborne Division, it included the 1st and 2d Battalions, 504th Infantry; 2d Battalion, 505th Infantry; 3d Battalion, 319th Field Artillery; and the 307th Engineer Battalion. Located at Kandahar airfield with smaller units stationed at forward operating bases in areas where the enemy remained active, the new task force conducted operations until it rotated out in late spring 2003.

In January 2003, Company A of 2d Battalion, 504th Infantry, and Company A, 307th Engineer Battalion, along with a contingent of coalition and Afghan forces conducted Operation MONGOOSE, engaging Hizb-e-Islami Gulbuddin fighters in the Hade Ghar Mountains near the city of Spin Boldak in eastern Kandahar Province. After heavy fighting, which required several air strikes by Norwegian F–16 aircraft, coalition forces entered an extensive underground complex filled with arms caches and documents, as well as food, water, and livestock. It appeared that the compound had been used to smuggle weapons and supplies from Pakistan into Afghanistan.

In June 2003, Special Forces teams set up small bases along the eastern borders of Nangarhar and Kunar Provinces to stop enemy fighters in Afghanistan from contesting Pakistani Army operations against al-Qaeda. Although the undetected movement across the border of armed insurgents using old smuggling trails remained a problem, the operation expanded the sparse coalition presence in northeastern Afghanistan. The deployment also improved cooperation with Pakistan, drew a larger Pakistani military presence to the border in that area, and, in U.S.-led talks, helped set the conditions for resolving border disputes between Afghanistan and Pakistan that became known as the tripartite meetings.

Relations between the United States and Pakistan, however, remained strained, with most of the tension stemming from the complex nature of Pakistani politics. Pakistan continued to allow U.S. supplies to move overland from Karachi into Afghanistan. President Pervez Musharraf had come out strongly against Muslim extremists in January 2002, banning several organizations from operating within Pakistan. In March 2003, Pakistani intelligence operatives captured Khalid Sheikh Mohammed, one of the principal architects of the 11 September 2001 attacks, and turned him over to the United States. In June, President Bush rewarded Pakistan with the promise of $3 billion in aid to be distributed over the next five years. However, Musharraf's crackdown on militant groups also led to two failed assassination attempts in late 2003. Caught between the United States and Muslim extremists, the Pakistani regime's behavior appeared inconsistent and duplicitous.

Dealing with Detainees

The question of handling captured al-Qaeda, Taliban, and Hizb-e-Islami Gulbuddin fighters became more pressing in the wake of escalating American-led combat operations. Heretofore, U.S. soldiers who detained enemy fighters operated under customary guidance to "search, segregate, silence, safeguard, and speed to the rear." Yet standard doctrine did not apply in all circumstances given the unique peculiarities of Afghanistan and the changing nature of the conflict.

Until late 2001, the simplest course of action consisted of turning over prisoners to the Afghans, but the lack of a centralized government in Kabul forced the Americans to rely on Northern Alliance warlords to oversee them. Events surrounding a prisoner uprising in late 2001 provided ample evidence of the problems pursuing this path. During the first month of the war, forces under Northern Alliance General Abdul Rashid Dostum assembled hundreds of prisoners in Qala-i Jangi, a nineteenth-century fortress near the city of Mazar-e Sharif. Following Afghan cultural norms, the captives were held in loose confinement after promising not to fight anymore against coalition forces. However, many of the prisoners were non-Afghan members of the Taliban or al-Qaeda. Ignoring their earlier promise, these prisoners mounted an uprising on 25 November that resulted in the deaths of two American intelligence officers. The ensuing battle to retake the prison facilities required the commitment of Dostum's troops, U.S. Special Forces, and coalition airpower.

Controversy arose regarding the subsequent incarceration and transport of the surviving prisoners by Dostum's forces. Numerous prisoners died while being transported to other facilities in sealed shipping containers, which resulted in media claims that American forces were somehow complicit in the deaths. While the actual number of deaths remained uncertain, no credible evidence emerged regarding possible U.S. involvement or knowledge of the event. Nevertheless, accusations of brutality levied against a key regional ally reflected poorly on the United States with Afghans, internationally, and with the American public. To avert similar controversies in the future, the Americans recognized they had to assume a more active role in detainee operations.

The U.S. decision to take responsibility raised a number of cultural and religious issues. For example, it is standard U.S. Army practice when dealing with lice among detainees to trim prisoners' hair and beards. For most Afghan Muslims, however, a clean shaven head and face served as a mark of shame. Should the prisoners be released, both they and their families would harbor ill will against coalition forces for the insult. Detainee prayers also initially caused some confusion at first. The Muslim practice of praying aloud caused some soldiers to think that the detainees were violating rules against talking with each other. Gender also created a point of friction. Male detainees reacted poorly to the military police women guarding them. As Pvt. Marisa A. Wondra of the 65th Military Police Company described it, "At first, they would think twice about what I'm telling them to do. 'Do I have to listen to the female?' But then they probably realized that I have a weapon and that they have to listen to me."

While cultural and religious differences proved troublesome, the most basic problems revolved around facilities and personnel. At the beginning of the conflict, adequate prison facilities were virtually nonexistent in a nation devastated by decades of war (and whose culture reflected little concern for prisoners in any case). Military engineers constructed a short-term detainee holding compound at Kandahar airfield that had guard towers, chain-link fence with razor wire, cells, showers, and latrines. A more permanent structure at Kandahar was eventually built, along with a larger detention facility at Bagram Air Base and smaller holding facilities elsewhere. Though austere at the start, the facilities improved over time. Authorities transferred select detainees to Guantanamo Bay, Cuba, and others were released outright or turned over to the nascent Afghan government, but enough remained in Afghanistan detention centers to cause difficulties for the Army.

Handling prisoners and intelligence gathering were the most critical elements of detainee operations. Military police internment and resettlement battalions, composed of an escort company, a guard company, processing detachments, and a headquarters unit, operated the detainee facilities. However, these specialized units were not always available in sufficient strength, with Guantanamo Bay and eventually Iraq having higher priorities than Afghanistan. As a result, units often undertook duties beyond the scope of their training, such as military police companies trained for escort duty being used to operate detention facilities. Furthermore, most internment and resettlement units came from the Army Reserve and National Guard, which created additional problems in training. Theoretically, their normal training regimen would prepare them for duty in Afghanistan. However, while they underwent predeployment training for service in a combat zone, many reservists received only cursory specialized instruction in detainee operations. Sgt. Alan J. Driver Jr., 377th Military Police Company, later charged

A detainee facility in Kandahar, February 2002

with abusing detainees, stated, "We received one day of training on prison techniques before we arrived in the country and actually took over the prison in Bagram." Lt. Col. Thomas S. Berg, a staff officer who investigated charges linked to Driver's unit, reported that "little of the training focused on the actual mission that the 377th anticipated it would assume upon arrival in theater."

The need to gather intelligence also complicated the detainee situation. Detainees were considered a crucial source of information, particularly since the coalition forces in Afghanistan were opposed by a clandestine foe hiding in the population rather than a conventional opponent. While military intelligence units expected to handle these interrogations, Special Operations Forces and other civilian agencies also questioned detainees, making it difficult to create any kind of procedural uniformity. The Defense Department and Central Command initially relied on Field Manual (FM) 34–52, *Intelligence Interrogation*, published in 1992, to provide basic guidelines for military intelligence personnel. However, some interrogators in Afghanistan began to use enhanced interrogation techniques authorized for Guantanamo Bay but not by FM 34–52, such as removal of clothing, prolonged periods of isolation, sleep deprivation, and stress positions. How these techniques migrated to Afghanistan remains unclear, but they contributed to uncertainty over allowable interrogation methods.

Frequent adjustments to detainee policies caused additional confusion, and it would be two years before a uniform policy appeared.

Army doctrine also gave conflicting guidance on the relationship between military intelligence personnel and military police. FM 34–52 implied an active role for military police in the interrogation process. However, FM 3–19.40, *Military Police Internment/Resettlement Operations*, published in 2001, stated that only military intelligence personnel could conduct interrogations at detention sites. It provided no rules for military police to assist in interrogations, nor did it spell out approved or prohibited military intelligence procedures at a facility operated by military police. In the absence of official guidelines, local commanders and their troops had to improvise.

The ambiguity between military intelligence and military police doctrine began with the classification of detainees. The lack of a formal declaration of war and a uniformed enemy raised the question of how to determine whether any given individual was in fact a fighter or involved in terrorism. Those captured did not readily fit under the Geneva Conventions rubric of enemy prisoners of war. Ultimately, they came to be labeled as *detainees* or *persons under control*. President Bush stated in early 2002 that prisoners would be treated in a humane manner, even though they were not accorded protection by the Geneva Conventions. But the administration and Defense Department provided no additional guidance as to how to deal with captives under this new classification.

With inadequately trained personnel operating under ambiguous guidelines, abuses occurred. Offenses included denial of access to medical care, repeated physical abuse, prolonged periods of detainment in shackles suspended from a ceiling, and the negligent homicide of at least two detainees. The most serious abuses occurred at the detention facility at Bagram Air Base, but cases of mistreatment took place at facilities throughout the country during 2002 and 2003. As a result of these abuses, a number of Army military police and military intelligence personnel received administrative punishments or faced criminal charges.

The problems in detainment operations in Afghanistan remained outside the public's knowledge until a scandal erupted in late April 2004 over claims of abuse at Abu Ghraib Prison in Iraq, prompting a review of all detainee practices in Iraq and Afghanistan. U.S. forces conducted a "top to bottom" evaluation of all coalition detention and holding facilities to make sure that procedures were "in accordance with the spirit of the Geneva Conventions." The study did not uncover systemic abuse but did see a need for better guidance and training for prison staff. Responding to criticisms, the Defense Department ended the practice of holding detainees indefinitely. New rules stated that prisoners being held at a temporary facility had to be released or transferred if they were held for more than ten days. All prisoners would have access to the Red Cross/Red Crescent within fifteen days of detention. A coalition spokesperson in December 2004 concluded that "U.S. forces have tightened up procedures for training up our people to handle and care for the prisoners." However, organizations such as Human Rights Watch disputed these official pronouncements and remained highly critical of the Army's oversight of detainees.

Creating the Afghan National Army

The coalition started planning for an Afghan army in January 2002, but that process required more time and effort than creating an Afghan national government. Afghanistan did not have the professional military tradition or capabilities to establish and sustain such a force. Though the nation had been at war for decades, the contenders were primarily militias, loyal to a particular warlord, or mujahideen guerrilla fighters. Neither group could provide a solid foundation for a professional army. The warlords served as a stopgap in the military sphere as they had in the political realm, with their militias continuing to fight alongside coalition forces. But the lack of broad ethnic representation and the politicized leadership of the militias made them poor representatives of national authority. The linguistic differences in the country, when combined with the absence of common procedures among the coalition to train the Afghan military, were major impediments to molding a cohesive national army.

In April 2002, the United States and its allies convened in Geneva, Switzerland, to discuss security issues in Afghanistan. The allies agreed on five "pillars" of reform, with a different nation taking the lead in each. The United States would create and train the Afghan army, while Germany would build the Afghan police. Italy accepted responsibility for reform of the justice system. The United Kingdom took on the counternarcotics problem, and Japan became the lead nation on disarmament, demilitarization, and reintegration of the Afghan warlords and militias.

The International Security Assistance Force under British command trained the first battalion of Afghan infantry. When Turkey took over leadership of ISAF in June 2002, it created a second battalion. In the spring of 2002, U.S. Special Forces units partnered with the French to begin training additional Afghan infantry units. Progress was slow.

That fall, the United States established the Office of Military Cooperation–Afghanistan to manage the overall effort to create the Afghan National Army. Normally an office of military cooperation is part of the U.S. Embassy and facilitates the sale or transfer of American equipment and the provision of technical advice to the local military. In this instance, the new organization's much broader mandate resulted in the group working under dual chains of command. For traditional equipping and advising issues, the commander coordinated with the U.S. ambassador. The Office of Military Cooperation–Afghanistan also received direction from CJTF-180 and its successors for overseeing the Afghan National Army's training and operational employment.

In October 2002, Maj. Gen. Karl W. Eikenberry took command of the Office of Military Cooperation–Afghanistan. His headquarters near the U.S. Embassy in Kabul started with just fifteen people, a not inconsiderable number given that similar efforts in Iraq a year later began with six soldiers. The staff grew slowly with augmentation from CJTF-180, Army reservists, and a number of civilian contractors. The Netherlands, Germany, Romania, France, and the United Kingdom also contributed people. The majority of the contractors worked for Military Professional Resources Incorporated, which hired retired officers to mentor senior Afghan leaders of the General Staff and the Ministry of Defense.

When Air Force Maj. Gen. Craig P. Weston succeeded Eikenberry on 30 November 2003, the staff still had only 90 out of an authorized 228 billets. Given the disparate source of personnel, the length of tours of duty varied, which resulted in heavy turnover in the office. While the Army typically sent someone for nine to twelve months, a marine served for only seven months, and Air Force personnel rotated every three or four months. In a single two-month period, the organization experienced an almost 100 percent turnover in staff. This contrasted with the rotation of combat units in Afghanistan and Iraq, which preserved some organizational continuity.

General Eikenberry designated three "top-down" priorities. First, he emphasized rebuilding the Ministry of Defense and the General Staff, so that these institutions could establish policy and operational direction. Second, he concentrated on increasing the size of the combat forces of the Afghan National Army. The initial phase of this development consisted of creating the Central Corps, a light infantry division numbering about 12,500 stationed around Kabul. From this starting point, the army would expand with four regional corps based in the cities of Kandahar in the south, Herat in the west, Gardez in the east, and Mazar-e Sharif to the north. Third, the Office of Military Cooperation–Afghanistan would form sustaining institutions for the Afghan National Army, including commands for recruiting, education, training, medical, acquisition and logistics, and communications and intelligence.

At first, coalition troops and U.S. Special Forces teams trained the Afghan soldiers while the Office of Military Cooperation–Afghanistan handled other matters. But when Operation IRAQI FREEDOM started in March 2003, special operations units could no longer be spared. Responsibility then fell to a new organization, Combined Joint Task Force PHOENIX, under Brig. Gen. F. Joseph Prasek and formed around Col. Mark A. Milley's 2d Brigade, 10th Mountain Division. The war in Iraq brought more change a few months later at the end of the brigade's tour of duty in Afghanistan. With regular combat forces covering two conflicts, the U.S. Army could not afford to continue assigning active duty units to the Afghan training mission. A contingent from the 45th Enhanced Separate Brigade of the Oklahoma National Guard under the command of Brig. Gen. Thomas P. Mancino assumed duties as Task Force PHOENIX II in the fall of 2003. Thereafter, each rotation of Task Force PHOENIX consisted of several hundred soldiers from a National Guard brigade augmented by coalition personnel and soldiers drawn from National Guard units from several states. The 76th Infantry Brigade of the Indiana National Guard assumed duties as Task Force PHOENIX III in August 2004. The 53d Infantry Brigade of the Florida National Guard took over as Task Force PHOENIX IV in 2005. Embedded teams from these commands mentored Afghan National Army units during their training and deployed with the units after graduation.

In December 2002, the international community and Afghan government met again at Bonn, Germany. All parties agreed that the new Afghan National Army would be a volunteer force numbering no more than 70,000, composed of 43,000 combat soldiers, 24,000 support troops, and 3,000 Ministry of Defense and General Staff personnel. In addition, each unit down to battalion would contain a mix reflecting the ethnic and regional makeup of the country. By fielding a national and representative force, the coalition leadership intended to send a message of Afghan national unity and demonstrate that the Kabul government could provide security for the population.

The challenge of forging an army from scratch proved daunting. Doing so in the midst of an ongoing conflict, which put a premium on fielding combat units at the expense of staffs and logistical troops, made the process even more difficult. In order to accelerate the formation of combat units, the plan for the Afghan National Army combined individual basic training with unit creation and unit training. Building on the initial method developed by the International Security Assistance Force, the program brought together recruits, organized them into a battalion (known as a kandak), installed a full array of unit leaders, and put the outfit through a training cycle that lasted about ten weeks. The greatest problem lay not with teaching basic skills to privates—though that proved hard enough in just ten weeks—but selecting and tutoring noncommissioned officers, junior officers, and battalion commanders. Although there were many veterans available from the militias and even the old Communist-era Afghan army, the requirement to balance ethnic backgrounds and ascertain loyalty made it problematic to fill out the leadership ranks of each battalion. In some cases, members of the same unit had fought on opposing sides in previous years. Beyond that, even generals and those in the Ministry of Defense had no experience with a reliable and respected national army. Their lifetimes had been spent in a system where ethnic and tribal ties were the rule and national-level security institutions were suspect at best. Every Afghan involved, from private soldier to the president, had to not only acquire new skills, but also develop a new mind-set.

A lack of adequate training facilities served to constrain the growth of the Afghan military. The Kabul Military Training Center on the eastern outskirts of the capital, established in 2002, could only be reached by a poorly maintained highway. The base did not have adequate sanitation, a decent mess hall, or heat in most of its dilapidated buildings. Trainers relied on cellular telephones with uneven coverage to speak with their superiors just a few miles away in the city. Meeting face to face, after a slow trip over inadequate roads, continued to be the most reliable way to talk. There was one geographic advantage at first, as all elements of the effort—the Ministry of Defense, the training center, and the Central Corps—were located in the environs of the capital. After watching recruits attempt to master Western military skills amid cold and squalor on a visit to the training center during the harsh winter of 2002–2003, General Eikenberry remarked, "This is the Valley Forge of the Afghan National Army."

On 30 August 2003, the formal activation of Central Corps, which reached a strength of ten thousand soldiers organized into three brigades, took place at Pul-e Charkhi, just east of Kabul. Efforts to field the unit at full strength had been thwarted by desertion. By February 2004, the Afghan National Army numbered only seven thousand, as soldiers continued to leave the ranks. Low pay accounted for many of those who left. As Mohammed Tahir, a platoon sergeant from Jalalabad, put it, "If we can't pay rent, we have to find another job." Arming and equipping the force proved just as difficult. The coalition provided financial support but drew the line at extensive funding for new hardware, particularly in a country awash in Soviet-era weapons. In addition to being plentiful, these weapons had the added advantage of being familiar to most Afghans, having been in use by the old army and the militias for decades. Many Eastern European countries were also willing to donate older materiel as they converted to more modern Western arms. The Mongolians contributed mobile training teams on artillery, Romanians gave instruction on T55 and T62 tanks, and Bulgarians sent training crews to teach how to operate armored personnel carriers. On the downside, the reliance on donated equipment prevented standardization in many areas, exacerbating problems with maintenance and spare parts.

Coalition partners contributed in other ways to the building of the Afghan National Army. The British created a program to train noncommissioned officers. The French established a command and general staff course. The Turks assisted in creating a National Military Academy in Kabul—the Afghan West Point. Italian teams trained the Afghan Central Corps on setting up vehicle checkpoints. The different styles among the coalition did not always fit together. For example, the French placed less reliance on noncommissioned officers, while the British and the Americans viewed them as the backbone of their military forces. Since the French played a major role in training Afghan officers, the latter developed an attitude about the use of noncommissioned officers that U.S. training teams had to work to overcome. Over time this problem diminished as Afghans familiar with American and British military culture assumed more instructor positions in the training and education programs.

In one of its first significant missions, the Central Corps provided security for the ratification of the new Afghan constitution in January 2004. The command also secured the opening ceremony for the Bagram-to-Kandahar section of the Ring Road, a loop connecting the major Afghan population centers of Kabul, Kandahar, Herat, and Mazar-e Sharif. The Central Corps did these missions with far fewer troops than authorized until Task Force PHOENIX completed training additional battalions in June. Thereafter, new units slowly moved out into surrounding provinces as they finished the cycle at the Kabul training base. The Afghan army now existed on more than just paper but still did not have the capability to take a significant role in combat.

Disarming, Demobilizing, and Reintegrating the Militias

As the Afghan National Army slowly formed, the process of disbanding warlord militias got under way. The program, formally titled Disarmament, Demobilization, and Reintegration, was administered by the United Nations under the leadership of Japan and managed by the Afghan Ministry of Defense, with assistance from other coalition forces. In the case of the warlords who had contributed to the Taliban overthrow, they could usually be integrated into the new government in some fashion. In a few instances, warlords straddled the fence, accepting the Karzai administration while keeping their military forces intact and maintaining their regional influence. Some also continued to participate in illegal smuggling and drug dealing.

In the autumn of 2003, two warlords vied for control over several provinces in northeastern Afghanistan. General Dostum, an ethnic Uzbek, controlled the town of Mazar-e Sharif and surrounding areas. His rival, Attah Mohammed, was an ethnic Tajik. Both men had tanks, artillery, and other heavy weapons. The two factions shelled each other near Mazar-e Sharif in early October, killing at least fifty fighters. British soldiers responding to the incident positioned their vehicles between the two forces to halt the exchange of fire. Minister of the Interior Ali Jalali and coalition military officials met with the leaders to negotiate a truce. A similar situation occurred in August 2004 between the governor of Herat, Ismail Khan, and militia commander Amanullah Khan. The Afghan government defused the situation by convincing Ismail Khan to accept a ministerial post in Kabul. The United Nations facilitated the process by banning warlords from running in the upcoming elections for president and parliament. As the presence of coalition troops reduced the potential for wielding power through armed force, and the prospects for a legitimate government increased, many militia leaders saw the handwriting on the wall and turned in their heavy weapons.

A British officer in the coalition headquarters, Maj. Gen. Peter Gilchrist, monitored the effort to make sure that Disarmament, Demobilization, and Reintegration remained honest and transparent. By May 2005, the warlords had relinquished most of their heavy weapons, with serviceable items helping to equip the Afghan National Army. The rest of the process did not always work as well. One officer in Afghanistan observed, "We used to call it two big 'Ds' with a little 'r' because any significant reintegration plan really wasn't there." In theory, reintegration provided former fighters with alternative skills, such as farming or car repair. Others could join the Afghan National Army. As of September 2004, about 13,200 Afghan men had availed themselves of one of these options, but estimates of the total number of militia were around 100,000, leaving a large number unemployed, still serving a warlord, or relying on freelance violence to make a living. Those Afghans who turned to banditry often formed into small bands and set up roadblocks in the countryside to extort illegal "tolls" from merchants and travelers.

Civil-Military Operations

The civil affairs mission in Afghanistan began on 13 December 2001 with the creation of the Combined Joint Civil-Military Operations Task Force in Kabul. Commanded by Army Brig. Gen. David E. Kratzer, the task force consisted of a mix of active and reserve units, to include elements of the 96th Civil Affairs Battalion, 122d Rear Operations Command, and 377th Theater Support Command. Kratzer's mission was to help establish good relationships with local authorities and the population, in large measure by bringing improvement to their communities. Decades of civil war, neglect, and isolation had left the majority of the country's rural areas severely underdeveloped, and the nation's anemic economy proved insufficient to generate the funds necessary for internal developments.

To achieve his mission, General Kratzer split the overall civil-military effort into two components. Company C, 96th Civil Affairs Battalion, based in Karshi-Khanabad, Uzbekistan, took responsibility for twelve northern Afghan provinces. Company D, operating from Kabul, covered seventeen provinces in the south, including the capital. Each company deployed teams of three to five members to each province within its area of responsibility. In an effort to emphasize their noncombat orientation, the teams were dubbed coalition humanitarian liaison cells (referred to as chicklets based on their acronym CHLC).

The civil affairs teams worked with coalition military forces, the United Nations, the interim Afghan government, U.S. government agencies, and nongovernmental organizations (NGOs) to evaluate humanitarian needs and coordinate assistance projects. In addition to dealing with poor roads, uncertain security, and inadequate numbers of personnel, the civil affairs teams found their efforts frustrated by lack of funds. At the onset, the Overseas Humanitarian Disaster and Civic Aid program, a high-profile Defense Department fund aiming for a quick impact, served as the only available source of money for these projects. But the time-consuming bureaucratic procedures that had to be followed to access those funds made rapid progress difficult, especially given the limited manpower available to the task force. Later, the civil affairs teams tapped into Commander's Emergency Response Program funds, a proven process, with fewer prerequisites, already used in Iraq, which allowed projects to get started faster. Lingering issues with roads, security, and lack of personnel would require more time and resources to address.

Beginning in March 2002, reserve component Special Forces and civil affairs units started to replace their active duty counterparts. Growing tensions with Iraq put a premium on these active duty units, preventing them from being committed to long-term deployments in Afghanistan. The 19th Special Forces Group, a National Guard unit from several states, replaced the 5th Special Forces Group. The 489th Civil Affairs Battalion, an Army Reserve unit from Tennessee, replaced the 96th Civil Affairs Battalion. At first, the 489th Civil Affairs Battalion did not deploy at full strength due to the belief that it would be in Afghanistan briefly, perhaps only until July. As the mission expanded, the rest of the unit arrived over the summer, along with augmentees from the 401st Civil Affairs Battalion from New York. In November, the 450th Civil Affairs Battalion from Riverdale, Maryland, took over the mission.

As the reserve civil affairs units began to arrive in theater, they discovered that CFLCC in distant Kuwait now authorized them to operate only in uncontested areas. For increased security and logistical support, the cells were also confined to the same locations as Special Forces teams. When CJTF-180 assumed command, its staff had not anticipated the wretched conditions in the country or the severity of the Afghans' humanitarian requirements. As reports on the desperate needs of the population made their way up the chain of command, CJTF-180 expanded the reach of the humanitarian cells, authorizing their deployment into more isolated areas. The 489th Civil Affairs Battalion soon established a permanent presence in eleven cities around the country. The soldiers often lived off the local economy, leasing vehicles, purchasing food, and paying for local labor. Enemy forces responded to the increased civil affairs presence by attacking the teams, which began taking security details with them when working among the population.

The 489th Civil Affairs Battalion oversaw the completion of some one hundred major projects, including the construction of schools, health clinics, hospitals, wells, and irrigation systems during its deployment. The largest project took place in Herat, where soldiers hired hundreds of local Afghans to dig out 166 miles of irrigation canals. School projects proved popular with the people. The battalion commander, Lt. Col. Roland de Marcellus, noted that so many students, especially girls, wanted an education that teachers had to hold classes in shifts. Civil affairs soldiers also distributed food and other humanitarian assistance. After an earthquake in Nahrin in the northeast, the civil-military task force organized the movement of forty-five tons of humanitarian supplies into the area within a 48-hour period via twenty-four Chinook helicopter sorties.

The humanitarian cells proved so successful that the program began to expand at the request of many provincial governors. But the small teams were still limited in what they could accomplish, often depending on international civilian organizations to undertake any long-term projects. The Afghan government and the coalition needed a means to do more, especially in areas not fully secure. At Central Command, civil affairs and special operations planners, as well as representatives from the State Department and the Agency for International Development, came up with a new idea. Over the summer and autumn of 2002, they developed a model for "joint regional teams." The CJTF-180's plans section improved on the initial concept, making final decisions on its structure in terms of force protection, the civil-military operations centers, and other details. The interagency effort resulted in the formation of robust teams consisting of sixty military and civilian personnel. The first joint teams, created using people already in Afghanistan, however, would not follow a rigid table of organization.

In December 2002, Joseph J. Collins, a retired Army colonel serving as deputy assistant secretary of defense for stability operations, announced a program to establish eight to ten of the new joint teams across Afghanistan. While their mission mirrored that of existing humanitarian cells, it extended further to encompass developing and strengthening local governance. The joint teams included security elements so that they could operate in volatile areas. In addition, they would have better access to quick reaction forces and close air support in case of hostile contact. General McNeill, President Karzai, and CENTCOM approved the final composition of the teams. President Karzai requested that the name be changed to provincial reconstruction team (PRT), which he thought emphasized the civil affairs role along with its connection to his government.

In January 2003, the first PRT went to work in Gardez in Paktiya Province on the border with Pakistan. Others colocated with combat units manning forward operating bases in key areas across the country, such as Bamyan just west of Kabul, Kunduz and Mazar-e Sharif in the north, Kandahar, and Herat, covering the main ethnic groups and former Taliban-held areas. Eventually, coalition partners fielded

A U.S. column moving over poor roads and rough terrain, December 2003

their own provincial reconstruction teams. The British took over the mission in Mazar-e Sharif and opened a new site in Maimanah near Turkmenistan. Germany replaced the U.S. PRT in Kunduz and started one in Faizabad in the far northeast. The New Zealanders assumed responsibility for Bamyan, and the Netherlands founded a team in Pul-e Khumri north of Kabul.

Each provincial reconstruction team's mission and composition differed based on local conditions. As the program developed, the standard organizational template consisted of about eighty people centered on two civil affairs teams (four soldiers each) and a force-protection platoon (forty soldiers) as well as support personnel, such as cooks, mechanics, and headquarters staff. There might also be military police, a psychological operations team, an explosive ordnance detachment, intelligence personnel, and medics. When available, a provincial reconstruction team also incorporated representatives from the Agency for International Development, the State Department, the U.S. Department of Agriculture, and the Afghan Ministry of the Interior. The majority of personnel and support for the PRTs came from the military because the armed forces had resources that civilian agencies lacked.

Command of the provincial reconstruction teams originally fell under the Combined Joint Civil-Military Operations Task Force. This arrangement led to significant issues whenever the teams operated in the same areas as combat troops and Special Forces detachments. Since each of these three elements reported through different chains of command, it proved difficult to determine which units were conducting operations in a given area. The CJTF-180 staff had to step in to synchronize missions and keep all three command chains informed.

As PRTs interacted with the local population, commanders observed a recurring pattern. At first, the Afghans appeared apprehensive about the introduction of the team. After the completion of several projects the initial uneasiness turned into curiosity and people started initiating contact with the team members. Eventually, the team developed a level of trust with the local population. One senior American general in Afghanistan characterized the appearance of the teams as "outposts of hope" in areas that previously had been visited by only armed soldiers seeking the enemy.

The commanders of the initial provincial reconstruction teams were told to be "creative" in lieu of formal training. As a result, the teams soon tackled a wide variety of tasks to include conducting village assessments, identifying projects, and monitoring humanitarian and reconstruction efforts. The early projects focused on "quick-impact" activities, such as digging wells, while providing other humanitarian-type assistance typical of the work done by aid agencies and contractors. Once the immediate needs of the population had been met, the teams moved into longer-term reconstruction projects, mentoring local officials and extending the reach of the Kabul government into the provinces. They also served a separate valuable function for President Karzai, providing another source of information that allowed him to better evaluate the situation outside Kabul.

Even as the PRTs grew more effective, the coalition's humanitarian efforts faced continuing challenges. One difficulty for the reconstruction team commanders was to manage the expectations of the local Afghan population. Although the teams had access to various sources of funding, needs far exceeded resources. Moreover, Afghanistan had few of the ingredients necessary to build on, such as an educated population, trained workers, and basic services such as electricity and running water. As one soldier later remarked, "The people say we're in there doing reconstruction. Well, there are days that I will tell you we're in there doing construction." She explained, "There's no power grid to repair because there was never a power grid to begin with." This problem repeated itself again and again as U.S. forces engaged in nation building in the truest sense.

Another issue that plagued the provincial reconstruction teams arose out of the fact that personalities and relationships played a large role in each PRT's success. The very nature of the teams—a mixture of military and civilians from multiple organizations—made it difficult to develop unit cohesion among people with different backgrounds and objectives. The challenges of working in a foreign culture were exacerbated in Afghanistan by the mix of ethnicities, ancient tribal and religious mores, decades of devastation and dislocation due to war, and antipathy toward outsiders. It took an enthusiastic and perceptive team commander to gain the cooperation of the local elders, the local government, and provincial governor to establish priorities and foster projects that were most wanted, most needed, and most likely to be effective. The division of command also made it vital that reconstruction teams collaborate with combat units operating in the same terrain, so that the teams could move more freely and gain access to more areas. Even under the best of circumstances, it was hard to judge the teams' effectiveness. They could complete small-scale relief projects and improve relations with various villages, but transforming those gains into solid support for the Afghan government and opposition to the Taliban remained uncertain.

Finally, the PRTs had trouble interacting with the numerous nongovernmental organizations operating in Afghanistan. Many of the latter had been active in the country for over a decade and took a cautious approach to ISAF and Operation Enduring Freedom forces' recent combination of military and humanitarian efforts. Several NGOs complained that special operations units, as well as members of other organizations such as the Federal Bureau of Investigation, the Central Intelligence Agency, and the Drug Enforcement Agency, often wore civilian clothes, which made it difficult for Afghans to distinguish between NGO personnel and similarly garbed coalition personnel. In response, General Kratzer announced in April 2002 that U.S. troops providing humanitarian assistance while in civilian clothes would wear identifying apparel. The new policy did not, however, apply to all American or international personnel operating in Afghanistan and only partially abated the concerns expressed by the NGO community.

The tension between NGOs and coalition forces highlighted the unique nature of the conflict. Afghan stability required a combination of military and humanitarian assistance in the midst of an active conflict. The need to foster and sustain a legitimate Afghan government sometimes blurred the lines between these areas, presenting a host of challenges to the United States and its international and Afghan allies. Unfortunately, while the United States continued to adjust its approach to Afghanistan, other events took center stage.

Reorganizing CJTF-180

Beginning in September 2002, the Bush administration began to increase diplomatic pressure on Iraq. Escalating diplomatic moves were mirrored by the deployment of expanding numbers of U.S. and allied troops to Kuwait. When Saddam Hussein failed to meet any U.S. deadlines or demands, 148,000 American, 47,000 British, 2,000 Australian, and troops from an assortment of coalition nations invaded Iraq with the purpose of overthrowing the Iraqi regime. Baghdad fell by 9 April 2003, prompting President Bush to declare an end to major combat operations on 1 May. Despite the relative ease of initial operations, the Iraq war evolved into a protracted struggle requiring the United States and its allies to maintain a significantly larger presence there than in Afghanistan. For Operation Enduring Freedom, the invasion of Iraq meant that the necessity to limit the number of troops and materiel deployed to Afghanistan grew more pressing. From 2003 to 2005, Operation Iraqi Freedom accounted for almost 90 percent of American forces deployed in the region, while Operation Enduring Freedom remained at just over 10 percent. The allocation of other resources reflected the proportion of troops in each country.

As most Americans focused on developments in Iraq in the spring of 2003, CENTCOM implemented another rotation of its forces and revision of its command structure in Afghanistan. Elements of the 10th Mountain Division, still under the command of General Hagenbeck, began rotating back into the theater in May to replace the 82d Airborne Division. Col. William B. "Burke" Garrett's 1st Brigade, 10th Mountain Division, formed the core of Task Force Warrior, which included 1st Battalion, 87th Infantry; 2d Battalion, 22d Infantry; and 1st Battalion, 501st Infantry, from the 172d Infantry Brigade based in Alaska. After a four-month stint with Task Force Phoenix training Afghan forces, 2d Battalion, 87th Infantry, joined Warrior in November to round out the new task force's ground maneuver forces. The task force also included 3d Squadron, 17th Cavalry; 3d Battalion, 6th Field Artillery; 10th Forward Support Battalion; 50th Signal Battalion; 126th Finance Battalion; and 3d Battalion, 229th Aviation. Completing Warrior were Battery A, 3d Battalion, 62d Air Defense Artillery;

Company A, 41st Engineer Battalion; and Company C, 5th Battalion, 159th Aviation. With roughly five thousand soldiers, Warrior began operations in August 2003 and remained in Afghanistan until April to May 2004 (*Chart 3*).

At the same time it rotated in new combat forces, Central Command decided to amalgamate the operational and tactical commands into a single echelon by eliminating Combined Joint Task Force–82 as a separate headquarters, reverting to the structure prior to creation of CJTF-180. The desire to avoid redundancies and improve efficiency provided the rationale. This enabled CENTCOM to maintain what it determined to be adequate forces in Afghanistan while still keeping the bulk of its resources available for impending operations in Iraq. For those in CJTF-180, it became evident that Afghanistan would remain an economy of force campaign in the larger war against terrorism.

In flattening out the command structure in Afghanistan, CENTCOM melded the staff of 10th Mountain Division into CJTF-180 when it worked alongside General McNeill's XVIII Airborne Corps command echelon until the latter rotated out in the fall of 2003. General Hagenbeck became the deputy task force commander, while General Vines replaced McNeill as the commanding general on 27 May 2003. It took some time to shake out the new arrangement. The 10th Mountain Division had received little advance warning of its new role as a joint command and thus had little time to prepare for it. In addition, a significant portion of its headquarters consisted of individual augmentees, many of them reservists who joined the organization in theater and all of whom were unfamiliar with the command. Beyond that, the merger of two commands and two distinct missions generated its own challenges.

The reorganization and new personnel policies reflected the continuing difficulty Operation Iraqi Freedom posed for CJTF-180. Soldiers involved in Operation Enduring Freedom felt that as the two campaigns competed for resources, Afghanistan slipped to secondary status. Iraq became the priority for personnel, funding, materiel, leadership, and intelligence resources. CJTF-180's requests for additional personnel were not approved by CENTCOM. Most of the Special Operations Forces, which had played such a vital role since October 2001, were diverted to Iraq. The international community tried to offset the shortfall by expanding its role in Afghanistan. On 11 August 2003, the North Atlantic Treaty Organization (NATO) formally took command of the International Security Assistance

Chart 3—Combined Joint Task Force–180: Second Rotation, May–October 2003

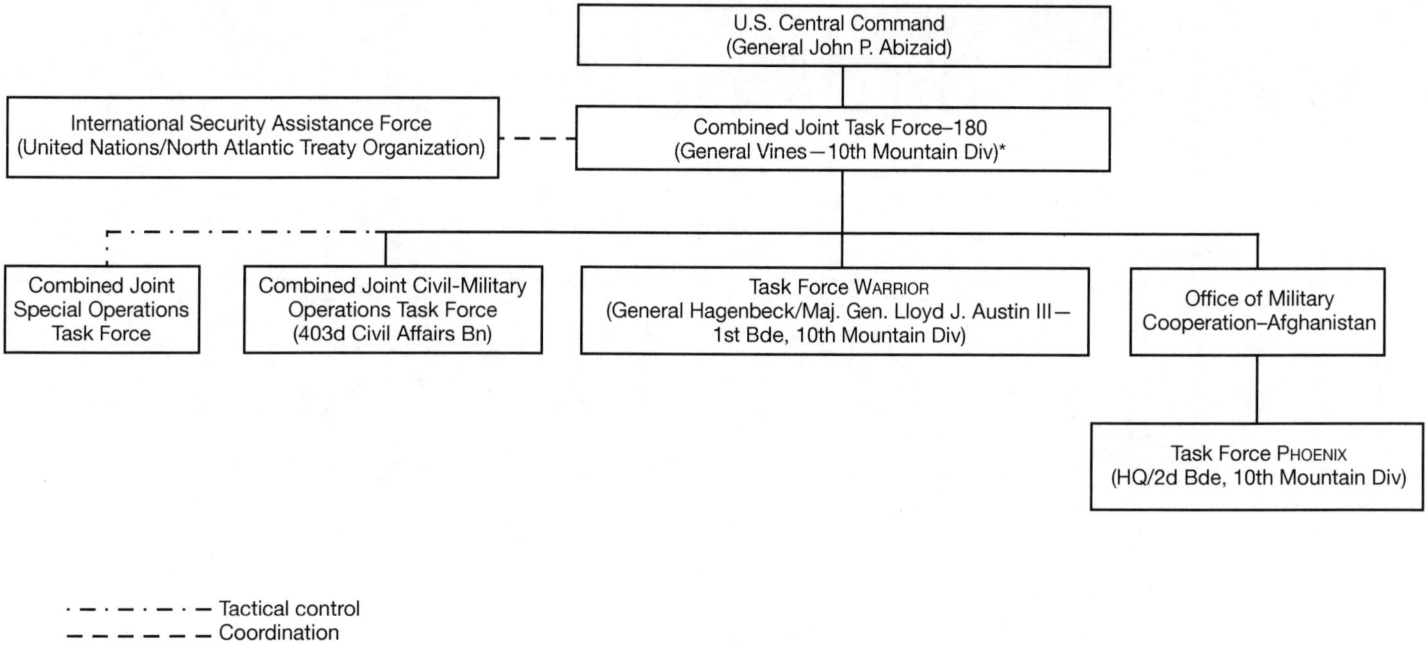

*From May to October 2003, the XVIII Airborne Corps and 10th Mountain Division staffs merged to form the Combined Joint Task Force–180 staff until the former rotated out.

Force, which had been providing security in Kabul. In October 2004, the NATO-led command assumed responsibility for nine northern Afghan provinces. That allowed U.S. forces to concentrate on eastern and southern Afghanistan, where a majority of the insurgents were located. However, no combined command over both ISAF and CJTF-180 existed, and the latter reserved the right to operate anywhere in the country. The operational limitations that the European governments placed on their contingents also made it necessary that U.S. forces be ready to deploy wherever a threat arose.

Even with these issues, U.S. forces generally worked hand in hand with their coalition partners and the Afghans. In July through September 2003, coalition forces conducted Operation Warrior Sweep, targeting the infiltration of al-Qaeda fighters from Pakistan into Helmand, Kandahar, Uruzgan, and Zabul Provinces. A combined force of 82d Airborne Division troopers in Task Force Devil, U.S. Special Forces detachments, Italian troops in Task Force Nibbio, and four battalions of the Afghan National Army worked to cordon off and search suspected enemy bases and confiscate weapons. Beginning in late July, helicopters inserted American paratroopers into the Ayub Khel valley in Paktiya Province at heights of ten thousand feet, so they could set up blocking positions. The paratroopers did not encounter enemy forces, but, acting on new intelligence that al-Qaeda operatives were in a nearby village, they moved to intercept them. With heavy packs, the soldiers hiked six miles over rough ground on narrow goat trails to reach the target. Once there, they uncovered caches of ammunition and explosives and captured a few suspected insurgents, but most had slipped away.

Although coalition and Afghan units augmented U.S. military efforts, a shortage of certain critical skills required U.S. units to use manpower in nontraditional ways. For example, explosive ordnance disposal detachments accompanied infantry units conducting dismounted patrols. The ordnance disposal soldiers dealt with mines, improvised explosive devices, and weapons caches along the way instead of remaining at the forward operating base until called on. One team leader, Sgt. Brett A. Fisher, recalled "humping in the mountains, carrying the packs, walking with them, doing these missions that technically we would not do in the States." He remembered a weapons cache his team discovered containing six tunnels bored a hundred feet down and full of shells and bullets. Most of the ammunition proved unserviceable, but the Ministry of Defense claimed the remainder for the Afghan National Army. The work could be extremely dangerous.

An example of a seized weapons cache

On 29 January 2004, a weapons cache exploded just west of the city of Ghazni, killing eight U.S. soldiers. It was the largest loss of American lives since Operation ANACONDA in March 2002. Although the coalition swept up tons of weapons and ammunition in the Bagram and Kandahar areas, the efforts had minimal impact on enemy stocks. As Brig. Gen. Benjamin R. Mixon, chief of staff at CJTF-180, observed, "Frankly speaking, this country is over running with those types of things, so I would suspect that we have only scratched the surface of the amount that is hidden."

During the course of 2003, it also became apparent to CJTF-180 that the Taliban sought to isolate Kandahar from the capital by taking advantage of long-standing tribal rivalries. While President Karzai came from the Pashtun Durrani clan, Taliban leader Mullah Omar was a member of the Pashtun Ghilzai. In addition to exploiting his natural ties to the Ghilzai, Mullah Omar stoked perceptions that Karzai had been favoring his own tribe. General Vines noted in September that as many as one thousand Taliban fighters were believed to be operating in Zabul Province, northeast of Kandahar. To help counter these efforts, the coalition headquarters directed subordinate echelons to place more emphasis on information operations and civil-military cooperation. As the Afghan National Army grew in size, Vines also encouraged the incorporation of more indigenous forces into combat operations.

With the 10th Mountain Division completing its rotation into Afghanistan, CJTF-180 devised a series of missions dubbed the mountain operations. Operation MOUNTAIN VIPER began on 25 August 2003 with the goal of combating the growing Taliban insurgency in southern and southeastern Afghanistan. Troops from Task Force WARRIOR, Special Operations Forces, and the U.S. Air Force took part in the operation. CJTF-180 sent Special Forces detachments into the mountains north of Kandahar, where they expected to find a small advance party of Taliban fighters. Instead, the teams discovered hundreds of insurgents gathered there. In response, a Special Forces unit and one platoon from Task Force COURAGE—the 2d Battalion, 22d Infantry—set up blocking positions in the valley near De Daychopan, a district north of Kandahar, as the rest of the task force conducted an air assault to flush out the enemy. The Americans expected the enemy to retreat into the mountains without a fight. Instead, the Taliban stood and fought for the first time since March 2002. U.S. forces routed the enemy, killing one hundred forty to two hundred insurgents.

CJTF-180 envisioned a follow-up operation, dubbed Mountain Viper II, to continue the pursuit. But the joint task force received guidance from CENTCOM to focus instead on Kunar and Nuristan Provinces in northeastern Afghanistan. Intelligence sources indicated that several al-Qaeda and Hizb-e-Islami Gulbuddin leaders might be there, including Gulbuddin Hekmatyar. In response to the information, soldiers conducted an air assault into a river valley in Nuristan Province within ninety-six hours, keeping pressure on enemy forces and denying them stable access to the area.

Operation Mountain Resolve began on 6 November 2003. One participant, Sgt. Carl A. Ashmead of Company B, 2d Battalion, 22d Infantry, explained, "Our job was to helicopter into the base of the valley and move up it, and basically flush any ACM [anticoalition military] out left, right, and forward, into these JSOAs [joint special operations areas] where the special operators would get them." The heliborne movement occurred at night into a zone secured by Task Force Catamount, the 2d Battalion, 87th Infantry. The landing zone was a terraced field, which caused problems for the all-terrain vehicles the soldiers brought to haul supplies. The men endured slippery conditions in the rain, while walking narrow goat trails wide enough for only a single person. The troops traveled across about thirty miles of mountainous terrain in five days to reach their objectives. The operation showed that American soldiers were becoming accustomed to the physical requirements that Afghanistan demanded. One officer explained,

> Soldier endurance was a highlight throughout the operation. The average soldier walked approximately 50 kilometers with a 100-pound load over the worst terrain that Afghanistan has to offer. He then conducted night security, search and attack operations, and patrols over a six-day period without fail. . . . They did this with supplies of one MRE [meals, ready to eat] and as little as three quarts of water in extreme temperatures of heat and cold.

After the soldiers climbed two thousand feet to reach the villages overlooking the valley, companies conducted search operations in nearby towns, interviewing residents for information on enemy activities in the area and seizing weapons caches and components for improvised explosive devices. Though Mountain Resolve did not inflict significant casualties on the enemy, the soldiers achieved their objectives of denying the area as a winter base of operations to the enemy and forcing enemy fighters to remain on the move when they were more likely to be captured or killed.

A gun position in an Afghan village, December 2003

Combined Forces Command–Afghanistan Takes Over

Over the summer and autumn of 2003, the stress of having a single headquarters responsible for the full spectrum of command and control, from the tactical to the strategic, again became apparent. CJTF-180's attention span ranged from small-unit skirmishes in the mountains to dealings with President Karzai and the leaders of surrounding nations. With the bulk of its staff coming from Hagenbeck's headquarters, CJTF focused on what it knew best—the tactical level. The departure of the remaining XVIII Airborne Corps staff in the fall exacerbated the problem. Brig. Gen. Lloyd J. Austin III, the assistant division commander (maneuver) for the 3d Infantry Division in Iraq until June 2003, arrived in September to replace General Hagenbeck as 10th Mountain Division commander. When General Vines departed in October 2003 for medical reasons, Austin assumed temporary command of all operations in Afghanistan.

In the late summer of 2003, General John P. Abizaid, who took over Central Command from General Franks in July, asked Lt. Gen. David W. Barno to establish a senior-level headquarters in Afghanistan. Designated Combined Forces Command–Afghanistan (CFC-A), its mission was to take over politico-military functions, build good relationships with both the U.S. Embassy and the Afghan national government, and coordinate with the International Security Assistance Force. CJTF-180 would focus on tactical operations, performing once again as a normal division headquarters.

Establishing CFC-A posed several challenges. General Barno arrived in November 2003 and set up his headquarters in Kabul within the U.S. Embassy compound with only six people. This initial attempt to operate with a skeletal staff, while relying on CJTF-180 at Bagram Air Base for further support, proved unsuccessful. Barno described this minimalist approach to creating a new command echelon as "the instant coffee model. Just add water and it springs out of the ground." The only saving grace came from the fact that half of the officers had served in his battalion in the 82d Airborne Division a decade earlier. That familiarity allowed them to "cut through all kinds of bureaucratic inertia in all the getting to know you and forming and storming things because we already knew and trusted each other and understood how each other thought and operated." Realizing that he needed to build a full-fledged headquarters, Barno began drawing personnel from CJTF-180, which in turn placed additional strains on that headquarters.

The CFC-A staff soon drafted a document that outlined manpower requirements for the headquarters, but the process of obtaining personnel proved difficult. Because CFC-A headquarters was competing with Iraq and its own combat forces for qualified active duty personnel, the Defense Department decided to staff the new headquarters with individual ready reservists, some of whom had not been on active duty in years. "I couldn't even get SAMS—School of Advanced Military Studies—graduates," Barno stated, "and I even asked the chief of staff personally for that at one point in time with no results." Meanwhile, the Army's peacetime personnel system continued to send those same graduates to units deployed elsewhere. CFC-A tried to rely on coalition partners to make up the shortfalls, designating as many as one hundred staff positions for coalition people, but other nations only filled twenty-five to thirty billets at a time. About four to five months after Joint Forces Command validated CFC-A's requirements in late 2004, the U.S. military services began sending quality personnel in sufficient numbers. However, it took almost eighteen months to fill out the headquarters. By May 2005, over four hundred soldiers staffed the command.

General Barno believed that both the process of creating a combined and joint organization and the quality of available staff would have been much improved had a standing Army headquarters been tasked to serve as the nucleus. On the other hand, he acknowledged the limitations of the one-year-rotation policy on any type of unit deployed to Afghanistan. "Every twelve months, it's the first day of school. . . . Everybody's trying to find their lunch box and get their rain jacket and discover where their shoes are, when we already knew all those things six months earlier." While a standing Army headquarters would have to cope with the frictions caused by rotation, the procedure used to fill Barno's staff compounded that problem by assigning individuals, instead of a cohesive group, to his staff.

Another issue for the new headquarters concerned security restrictions on coalition personnel. Americans conducted their work on a classified Internet system and often accessed documents classified as secret/NOFORN, which meant the information could not be released to foreign nationals. Although the coalition staff had use of a parallel system, Americans often did not transfer material to it. This posed problems both within the headquarters and with subordinate commands. An Italian unit, for example, could not receive intelligence despite operating with U.S. forces. Sometimes the results bordered on the absurd, except for the very real impact. In one case, a British officer could not read intelligence designated NOFORN, even though the original material had come to the United States from the United Kingdom. Only in May 2005 did key coalition staff receive restricted access to the secure American Internet network.

Shift in Mission Focus

One of General Barno's first priorities was reevaluating the operational situation. As Barno saw it, the information available to him slanted toward a military view of the situation, often described as kinetic operations. Units made extensive reports of combat activities but did not have the expertise to assess the political and economic status of the areas in which they worked. What the commander of CFC-A really wanted to know was "the nuance there, what was the environment, what was the sense, and what was happening." Barno could get information on combat operations, but he could not find out who was the chief of a particular village, whose side the chief preferred at the moment, and what opportunities there were to better the lives of his people. Barno found that the best information came from the PRTs' State Department representatives, who spent extended periods in a specific region collecting intelligence and building relationships with the people. Barno believed that for the coalition to be successful its tactical units needed to develop these same connections with the Afghan people.

After reviewing the situation on the ground, Barno determined that Afghanistan "was at a point of transition—a strategic fork in the road." He concluded that the initial post–11 September strategy, which focused on combating terrorists and their supporters, seemed less applicable in the wake of growing efforts by the Taliban to topple Karzai from power. Barno consequently called for a change in coalition military emphasis from counterterrorism to counterinsurgency. In addition to combat operations, the coalition needed to help the Kabul government protect the Afghan people while persuading them to join the fight against the enemy. He summarized the new course: "My focus was on the people of Afghanistan. The people of Afghanistan were the center of gravity."

At the time, the U.S. Army did not have an updated doctrine for this type of campaign, and Barno admitted, "None of us really had much of any training on the counterinsurgency business." To fill in the slate, he went back to basics, starting with his 1974-edition history textbook from the United States Military Academy at West Point, which described counterrevolutionary warfare. Some Army units ordered books on the subject, including the Marine Corps' *Small Wars Manual*, first published in 1935, and John Nagl's *Learning to Eat Soup with a Knife*, printed in 2002. General Barno also sought a British deputy since he believed that nation had considerable counterinsurgency experience. However, not everyone felt disadvantaged with the lack of official doctrine. The chief of staff of CFC-A, Col. David W. Lamm, believed "we were blessed by the absence of existing doctrine" because that allowed initiative to flourish.

The changing mission focus necessitated a fundamental shift in operations and unit deployments. Previously units had sallied forth from bases to search for the enemy in a designated location selected as a result of available intelligence. They might sweep one region in a given week and helicopter into an entirely different one far away the next month. Now General Barno assigned each unit responsibility for an area of operations—for the most part aligned with the political boundaries of provinces and districts—where it would operate for its entire tour in country. The switch not only allowed a company or battalion to become familiar with the terrain, but also made it possible for soldiers to develop relationships with the local leaders and people. And those ties permitted the coalition to tap a previously overlooked source of intelligence needed to wage a successful counterinsurgency campaign.

Combined Forces Command–Afghanistan summarized General Barno's new emphasis in a formal campaign plan designed to govern operations for the next five years. It consisted of five "pillars" of operation. Counterinsurgency operations constituted one pillar, with the effort focused on turning the population away from supporting the Taliban, al-Qaeda, or Hizb-e-Islami Gulbuddin. A second involved the familiar goal of building the Afghan National Army and the Afghan National Police. Establishing area ownership for coalition forces constituted the third, while the fourth repeated the earlier emphasis on building good governance and extending the reach of the Afghan central government into the provinces. Finally, General Barno himself would handle the fifth pillar, engaging the regional states: Afghanistan, Pakistan, Tajikistan, and Uzbekistan.

A critical feature that enabled General Barno's strategic shift and assignment of areas of operations was the decrease in violence during 2003. At the beginning of Task Force WARRIOR's deployment, it engaged primarily in direct combat with limited opportunity to build a relationship with the local population. Although the northern part of Afghanistan remained calm, violence often erupted in the southern and eastern areas of the country. A year later, conditions quieted down enough so that the troops could balance their time between combat operations and humanitarian assistance. With less than fourteen thousand troops in Afghanistan at the beginning of 2004, the coalition had to spread its forces very thin to maintain a visible presence in the Afghan provinces. In some instances, a battalion task force (roughly eight hundred soldiers) had responsibility for a 2,000-square-mile region. This dispersion required significantly more preparation by tactical units before they could perform large-scale counterinsurgency operations but did enable them to maintain a small but consistent presence throughout the provinces. Barno's goal was to create a level of comfort and trust between Afghan civilians and coalition forces that could be developed into positive relationships. By balancing offensive operations, humanitarian projects, and intelligence gathering, General Barno hoped to produce stable environments within the provinces that would not support renewed Taliban infiltration. At the same time, however, the low density of coalition troops meant that they would not be able to provide sustained security in the remote villages that defined Afghanistan.

One of the first operations conducted under the new campaign plan focused on furnishing security for the December 2003 loya jirga, which met in Kabul to agree on a constitution. Maj. Gen. Lloyd J. Austin III, the CJTF-180 commander, sought to prevent the Taliban from disrupting the gathering. As a result, CJTF-180 settled on Operation MOUNTAIN AVALANCHE as a means to pressure the enemy across a six-province zone. The plan called for Task Force WARRIOR to focus on Taliban forces operating in and around Kandahar while special operations units targeted Hizb-e-Islami Gulbuddin groups in Kunar Province along the Pakistan border. More than two thousand U.S. and Romanian troops conducted patrols and searches, uncovering several weapons caches that included tanks, armored personnel carriers, and artillery, as well as improvised explosive devices. Coalition forces also killed ten enemy fighters and detained over one hundred people suspected of ties to the insurgents, enabling the constitutional loya jirga to complete its work without disruption.

Task Force WARRIOR conducted two other major operations involving thousands of troops scouring the countryside during its rotation: MOUNTAIN BLIZZARD in January 2004 and MOUNTAIN STORM in March. These operations disrupted enemy lines of communications stretching between Zabul and Kandahar Provinces to safe havens in Pakistan. MOUNTAIN STORM, the final operation by WARRIOR, focused on establishing suitable security conditions for the Afghan presidential election scheduled for the fall of 2004. Coalition forces, including the 22d Marine Expeditionary Unit and Afghan National Army units, coordinated a hammer-and-anvil operation, where Pakistani troops drove the enemy into stationary ambushes set up by coalition troops in Afghanistan. Unfortunately for the coalition, the Pakistan Army proved less than capable. The movement of as many as twenty-five thousand Pakistani troops into North and South Waziristan led to significant casualties for those forces. The operations mirrored Pakistan's efforts in the Federally Administered Tribal Areas, where government forces launched a series of incursions in the spring of 2004. Though the efforts produced hundreds of enemy casualties, the Taliban and al-Qaeda leadership survived and maintained their presence in the region. The Pakistani offensives followed a sad pattern in which early enthusiasm quickly collapsed due to high casualties, forcing President Pervez Musharraf to seek truces with various tribes in Waziristan and the tribal areas. Circumstances had compelled the two sides to cooperate, but relations between Pakistan and the United States continued to be strained.

On the whole, however, U.S. efforts in southern and eastern Afghanistan followed a typical pattern by the spring of 2004. Following an operation to clear an area, civil affairs teams would come to deliver food, build a well, or offer other assistance. Medical teams provided care to villagers and their animals. Chaplains visited with local religious leaders. Psychological operations units transmitted radio programs and distributed leaflets to the population. The broadcasts did not reach everywhere, however, allowing the Taliban to spread propaganda almost unopposed in some areas.

Coalition casualties remained low during these operations in 2002 through 2004, with an average of five deaths per month (both combat and nonbattle). However, one loss stood out in the public eye. Patrick D. "Pat" Tillman had gained national attention in 2002 when he gave up a multimillion dollar National Football League contract with the Arizona Cardinals to enlist in the Army. By the spring of 2004, he and his brother Kevin were serving with the 2d Battalion, 75th Ranger Regiment. On 22 April, Private Tillman's platoon had been delayed in its mission to search a village by a broken-down high-mobility multipurpose wheeled vehicle. To complete the assignment before dusk, half the unit moved out toward the objective, while the rest followed with the disabled vehicle in tow. Shortly after departing, the lead element (including Tillman) lost communications with the slower group. Insurgents then ambushed the latter. Hearing the sound of firing, Tillman and others rushed back to the top of a nearby hill to provide cover. He and an Afghan soldier were killed by friendly troops who mistook them for enemy fighters.

The Army's initial public announcement claimed that Tillman died as a result of enemy action. He was awarded a Silver Star and promoted to corporal, even though some senior leaders had reason to doubt the veracity of initial reports. The Tillman family did not receive notification of the investigations or the final determination that he died due to friendly fire until thirty-five days after Tillman's death, despite Army regulations stating that next of kin be notified as new information became available. Members of Tillman's unit and chain of command violated Army protocols for processing and reporting the incident. Investigations brought the true story out, but not before causing additional pain to the Tillman family and chagrin for the Army. The affair embarrassed the Bush administration while providing critics of the war with additional ammunition.

In addition to the Tillman case, the changing nature of the conflict that accompanied the shift in mission focus was not always positive for the coalition. As the enemy grew to understand the rules of engagement, he exploited those restrictions. One of the regulations prohibited Army forces from crossing into Pakistan, so insurgents launched rockets or mortar rounds from Afghan territory before retreating over the border to safety. Another rule prohibited firing on unarmed persons. In one case, four Taliban ambushed a convoy with small arms and rocket-propelled grenades, prompting soldiers to call for air support. By the time helicopter gunships arrived, the fighters had discarded their weapons and were walking down a hill, preventing the pilots from engaging them. The ground convoy located and detained the men, but in the absence of weapons or other conclusive evidence, it was an open question how long they would remain prisoners. Events such as this prompted the chief of staff of CJTF-180, Brig. Gen. Byron S. Bagby, to give the Taliban forces their due: "They are a living, breathing, and thinking enemy."

Reorganizing Under Combined Joint Task Force–76

Between February and April 2004, the 25th Infantry Division from Schofield Barracks, Hawaii, replaced the 10th Mountain Division as the combined joint task force and provided a number of individual augmentees to Combined Forces Command–Afghanistan. In recognition of the transfer and in a nod to the year of American independence, CJTF-180 changed its designation to Combined Joint Task Force–76 (CJTF-76). The 25th Infantry Division commander, Maj. Gen. Eric T. Olson, assumed command of CJTF-76 on 15 April.

According to General Barno's plan for deploying coalition forces, CJTF-76 divided its area of responsibility into two main sectors, now called regional commands. On 6 May 2004, 3d Brigade, 25th Infantry Division, conducted a transfer of authority with 1st Brigade, 10th Mountain Division (Task Force WARRIOR), at Kandahar airfield. With the additional forces already at Kandahar, the 3d Brigade transformed into Combined Task Force BRONCO under the command of Col. Richard N. Pederson. It included the 1st Battalion, 501st Infantry; the Romanian 280th Infantry Battalion; the French Special Operations Task Group ARES; two Afghan National Army battalions; and four provincial reconstruction teams. Containing over three thousand soldiers and headquartered at Kandahar airfield, BRONCO had Regional Command South (RC South) as its area of responsibility, encompassing Nimroz, Helmand, Kandahar, Zabul, Uruzgan, Paktiya, Khost, Paktika, and Daykundi Provinces.

Responsibility for Regional Command East (RC East) originally fell to the 6th Marines. The area of operations included Bamyan, Parwan, Kapisa, Nuristan, Kunar, Laghman, Kabul, Wardak, Logar, Nangarhar, and Ghazni Provinces. The 25th Infantry Division Artillery headquarters, commanded by Col. Gary H. Cheek, took over RC East in June 2004. It formed Combined Task Force THUNDER, composed of 2d Battalion, 27th Infantry (with the addition of 3d Battalion, 116th Infantry, 29th Infantry Division, in July and the 1st Battalion, 505th Infantry, in September); 3d Battalion, 3d Marines; three Afghan National Army battalions; a military police platoon; eight provincial reconstruction teams, including a New Zealand reconstruction team; and Logistics Task Force 524. The primary mission for the four thousand soldiers in THUNDER was to stem the growing stream of weapons and Taliban recruits flowing into northeastern Afghanistan from Pakistan (*Chart 4*).

The 25th Division Artillery headquarters' deployment to Afghanistan was the first instance of a field artillery fires brigade deployed as a maneuver headquarters. It brought over only one battery—Battery F, 7th Field Artillery—which carried 120-mm. mortars instead of its standard 155-mm. howitzers. The battery returned to 155-mm. howitzers in August 2004. Because of the differences in the table of organization and equipment between an artillery and infantry brigade headquarters, the divisional artillery saw its staff grow to include all elements necessary to undertake a nonstandard mission.

The situation in Afghanistan's western provinces proved less tumultuous in early 2004. The 3d Squadron, 4th Cavalry, and Company B, 193d Aviation, initially took over responsibility for the area as Combined Task Force SABER. When tensions between rival warlords increased in September, CJTF-76 created a third area of operations, designated as Regional Command West (RC West), to monitor activities in Herat, Farah, Ghor, and Badghis Provinces. All but one of these provinces bordered Iran. Three Afghan National Army battalions and two provincial reconstruction teams were added to SABER, creating a 2,400-strong force, renamed Combined Task Force LONGHORN, under the command of Col. Phillip Bookert. Bookert's command would stabilize the region and pave the way for its eventual transfer to ISAF control.

Afghanistan presented a conceptual change for the 25th Infantry Division. Prior to its deployment, the division had concentrated on training for conventional operations. Once in Afghanistan, the division's officers had to rethink how they would operate to ensure the actions of their units supported Barno's focus on counterinsurgency. As Brig. Gen. Bernard S. Champoux, the assistant division commander (operations) for the 25th Division who was serving as the deputy commanding general of CJTF-76, explained,

> I think, initially, we didn't understand the insurgency. I think we thought we were going to go kill/capture and defeat the Taliban and al Qaeda

Chart 4—Combined Forces Command–Afghanistan: First Rotation, October 2003–April 2005

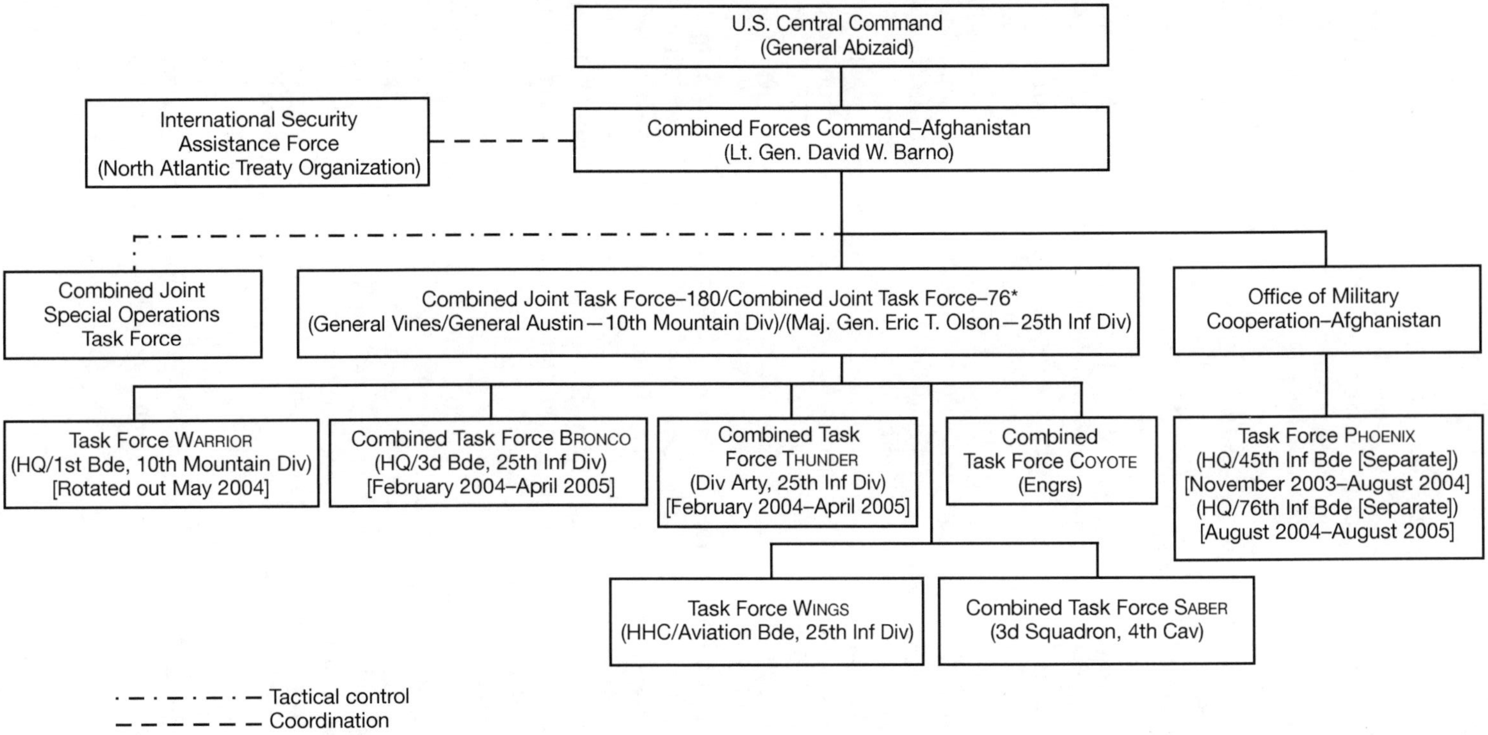

*This changeover occurred in April 2004.

> and Hekmatyar. In a short window of about sixty days, we really started rethinking all of those things. . . . Really, what we were confronted with was voter registration and this election, and then the inauguration and the constitution. Those are the things that became measures of success, not how many Taliban we killed and captured, or al Qaeda.

By late summer, the 25th Division's task forces were using the traditional "ink spot" or "oil stain" operating method when they could. Troops pushed out from the division's larger garrisons around Kandahar by establishing smaller forward operating bases in contested areas, initially the southern and eastern provinces. Once several smaller bases were established, the soldiers began conducting progressively overlapping patrols in order to deny the enemy access to the local population.

In July 2004, General Barno decided to transfer responsibility for Paktika, Khost, and Paktiya Provinces from RC South to RC East to balance out the territory between the regional commands. The combined task forces exercised operational control over every unit in their zone—ground, air, and logistics—with the exception of Special Operations Forces, which still came under the Combined Joint Special Operations Task Force. The Special Operations Forces elements were careful to coordinate with the brigade commanders whenever they worked in those zones. In addition to conducting combat operations, the brigade task forces mentored and worked with recently deployed Afghan security forces, fostered civil-military relations with the Afghan population, and cooperated with other government agencies in an attempt to increase popular support for the Afghan government.

General Barno decided that setting conditions for a successful Afghan presidential election would be the main effort in 2004. He requested and received additional troops for that mission, and by June U.S. forces numbered around eighteen thousand, up from just under ten thousand a year before. The International Security Assistance Force also grew in size. A Spanish light infantry battalion served as a quick reaction force stationed in Mazar-e Sharif, while the Italians added a battalion as the theater's reserve. In addition to fixed- and rotary-wing aviation units, NATO designated the 1st Battalion, 63d Armor, based in Vilseck, Germany, to serve as an on-call reserve if required by ISAF.

CJTF-76 dubbed the operation to secure the presidential election LIGHTNING RESOLVE. Coalition forces conducted joint security operations with the Afghan National Army, Afghan National Police, and the United Nations, as all parties prepared for the election. They also met with local elders, religious leaders, and officials to involve them in the election process and the security of the elections. Not surprisingly, insurgent forces attempted to undermine voter registration. In July 2004, an improvised explosive device planted in a mosque in Ghazni Province killed two United Nations employees registering voters. Other problems hampered election preparations. The Afghan government's limited infrastructure, the paucity of trained personnel, and security concerns throughout the country caused delays in voter registration. The slow rate of disarmament, demobilization, and reintegration also created fears within the coalition that the election could spark organized opposition. As a result of these concerns and the need to provide more time to register voters and disarm the militias, the Afghan interim government postponed the presidential election from June until September and then again until October, and the parliamentary elections were rescheduled for the following spring.

The focus on preventing Taliban interference in the elections also influenced the use of the Special Operations Forces. There were only twenty-five of its operational detachments alpha in the country at that time, totaling about two hundred fifty men. In April 2004, CJTF-76 directed the Combined Joint Special Operations Task Force to disrupt the movement of insurgents and materiel from Pakistan into Afghanistan under the code name Operation TICONDEROGA. The directive called for the special operations task force to shift assets from the interior of the country to the eastern border areas, in coordination with operations by the Pakistani military in Waziristan and the tribal areas. In Regional Commands East and South, Special Forces operated out of fire bases along the boundary with Pakistan. In order to move into Paktika Province, the Special Forces teams had to assault and reoccupy abandoned camps near Shkin and Lwara.

From 30 September until 10 October 2004, Special Operations Forces elements conducted Operation TRENTON to preempt enemy attempts to disrupt the elections. On 8 October, the day before the voting, about two hundred Taliban attacked a Special Forces base camp in the Deh Rawod District of Uruzgan Province. The teams returned fire and called in air support, killing about seventy fighters. The enemy also rocketed Special Operations Forces positions along the Pakistan border. The Americans responded with artillery, killing a number of insurgents. A short while later, another force of about twenty Taliban crossed the border into Zabul Province to link up with fighters already inside Afghanistan, but a Special Forces detachment directed bombing missions that destroyed the group.

Election day arrived on 9 October 2004. United Nations personnel had registered 10.8 million voters, about 42 percent of them women, and they turned out in large numbers to cast their ballots. In Bamyan, citizens lined up in the snow five hours ahead of the opening of the poll station. Although officials expected the Taliban to do their utmost to disrupt the process, only a handful of attacks took place with approximately one hundred sixty planned attacks thwarted. At the end of the day, coalition forces worked with the Afghans to move the balloting materials and secure the counting places. As expected, Hamid Karzai won the election. He took office in December, with coalition military units assisting Afghan security forces protect the ceremony.

Searching for Stability

After the presidential inauguration, U.S. and coalition forces launched several operations in RCs South and East, each lasting from three to ten days, in support of the next phase of the electoral process that called for parliamentary elections. Generally, if Army forces did not encounter resistance when entering an area they immediately transitioned to conducting civil affairs assessments and providing humanitarian aid. Despite the overall success of these operations, the parliamentary elections were again postponed until September 2005. The justification for the decision was to give the government more time to finalize logistics, such as producing a more accurate census and developing a clearer ballot. However, news reports pointed to growing security concerns as an alternative explanation.

The slow process of building up the Afghan National Army contributed to public fears regarding security for the parliamentary elections. Through the end of 2003, the coalition training regimen formed and trained one Afghan battalion at a time. With a ten-week cycle, that meant only about five new battalions a year. Secretary of Defense Rumsfeld asked the Office of Military Cooperation–Afghanistan to accelerate this process so the coalition could hand off responsibility for security to the Afghans more quickly. During 2004, the training system expanded to produce multiple battalions simultaneously. By March 2005, 18,300 Afghan soldiers had graduated (totaling roughly 31 battalions), or about 42 percent of the authorized end strength of 43,000 personnel.

While the expansion of the Afghan National Army gradually accelerated, it focused on infantry battalions, leaving their facilities and supporting logistical system underdeveloped. The army had quickly developed shortages in uniforms, serviceable equipment, weapons, vehicles, and habitable barracks, among other necessities. Afghan units deployed in the field lacked a functioning logistical system, so Task Force PHOENIX had to step in to fill the gap on an ad hoc basis. Accelerated instruction also taxed the training teams. Producing additional battalions required more teams, but increased staffing did not arrive quickly, forcing reassignment of personnel from other duties and a decrease in the size of teams from sixteen members to twelve.

Other factors also degraded the Afghan National Army's combat effectiveness. Without a banking system, Afghan soldiers periodically took leave to deliver money to their families. Frequently, they remained at home past their authorized leave time to help harvest a crop or meet other needs. These absences created uncertainty about the present-for-duty strength of Afghan units, as some soldiers returned after an extended delay while others never came back. During its first two years, the Afghan National Army's desertion rate averaged around 20 percent, dropping into the teens over the next two years. Therefore, while the Afghan National Army's overall performance slowly improved, the main task of providing security and humanitarian assistance and building support for the national government remained with the coalition.

As systemic problems continued to plague coalition efforts to create the Afghan National Army, General Barno strove to counteract any perceptions of a faltering local government by expanding the provincial reconstruction team program. Since there were too few civil affairs personnel to meet the need, he disbanded the Combined Joint Civil-Military Operations Task Force headquarters and dispersed most of its personnel to the new provincial reconstruction teams. By the spring of 2004, Combined Forces Command–Afghanistan had increased the number of teams operating around Afghanistan to twelve, with the newest ones in the southern part of the country where the Taliban proved most active. The dissolution of the civil-military task force had the added benefit of consolidating the chain of command, since operational control of the teams now resided with the regional commands. The transition proved difficult for some teams, with one team falling under six different commands in a six-month period.

By the summer of 2004, the relationship between provincial reconstruction teams and civilian agencies had matured, and both worked cooperatively to complete larger projects. In the city of Lashkar Gah in Helmand Province, the Agency for International Development funded school-building projects, while the Commander's Emergency Response Program provided money for books, desks, chairs, carpets,

Afghan women and children awaiting medical care, March 2004

and computers. In a similar fashion, the Agency for International Development built medical clinics, the World Bank arranged for staff and some equipment, and the provincial reconstruction team paid to refurbish existing clinics and distribute more medical equipment. The reconstruction team commander also funded police stations, road construction, and buildings for the Women's Association, among other projects.

For all the good the reconstruction teams did, they did not do as much as desired. In a culture where personal rapport was critically important, frequent turnover in personnel hampered good relationships with local leaders. The pressure to show quick results sometimes influenced the choice of projects or methods in ways that favored short-term outcomes over long-term progress. On occasion, structures and equipment were put in place even though the local population did not have the skills or money to maintain them on their own. The small size of the teams, the country's poor infrastructure, harsh terrain and weather, and the fragile security environment meant that reconstruction teams often had the ability to affect the lives of only a few Afghans in nearby areas. The provincial reconstruction team concept was sound, but there simply were not enough people, money, and time to realize its full potential. They remained selectively useful but insufficient to meet the country's huge needs.

A soldier hands out school supplies to Afghan children, March 2004.

The Calm Before the Storm

In the winter of 2004–2005, the Taliban went to ground, as irregular Afghan fighters had traditionally done for centuries. With the onset of the early spring thaw, the Taliban resumed operations. In February 2005, fighters launched three attacks against coalition forces in southeastern Afghanistan. They ambushed an Afghan National Army patrol in Helmand Province and detonated a roadside bomb and employed small arms against an American unit in Kandahar Province. The insurgents also attacked ground troops in Khost Province in order to fire on responding helicopters. But after years of adapting to the situation in Afghanistan, coalition forces responded to these new attacks by stepping up their own activity. A Romanian battalion, which had performed only base security for Kandahar airfield after the formation of Combined Task Force Bronco in May 2004, began conducting mounted patrols in RC South. Some Romanian patrols involved overnight stays in Uruzgan Province and eastern Helmand Province because the unit, with vehicles dating to the 1970s, could not operate effectively in the heat of the day. In another example, U.S. artillery batteries in RC South divided into two-gun sections and deployed into austere forward operating bases from which they could provide fire support to Special Operations Forces and other units. The artillerymen not only manned their pieces, but also served as infantry if called on to operate on foot beyond the fire bases.

In 2005, the Afghan National Army also began to take a wider role in combat operations. An Afghan battalion led Operation NAM DONG in Uruzgan Province in April, with Special Forces advisers assisting the unit in establishing a command post and directing fire missions. That same month, the 1st Kandak (Afghan battalion), 3d Brigade, conducted Operation MINESWEEPER with coalition forces around Herat and Shindand. The Afghan soldiers searched homes, uncovering weapons caches and capturing a suspected local anticoalition militia leader.

The increasing presence of Afghan troops contributed to the population's willingness to aid coalition forces. Between February 2004 and February 2005, the number of weapons caches reported by Afghans almost doubled and accounted for 43 percent of all such stocks uncovered by the coalition. The overall number of improvised explosive devices turned in by Afghans rose 30 percent from summer 2004 to spring 2005, totaling 90 percent of all roadside bomb discoveries. The quantity and quality of intelligence coming from local civilians, police, and military provided tangible evidence that U.S. and coalition forces were achieving some results in Afghanistan.

In an example of the overall expansion of coalition efforts in early 2005, Army engineers moved beyond base construction projects to take a greater role in civil development. Elements from the National Guard rebuilt a segment of the highway between Kandahar and Tarin Kot in Uruzgan Province, which formed part of the Ring Road. The Agency for International Development funded the project, and, though security concerns prevented civilians from building parts of it, American engineers trained their Afghan National Army counterparts on the project and also used local contractors when possible. Completion of that section of the highway reduced driving time between the two cities from eighteen hours to five. Afterward, Combined Task Force BRONCO troops began mentoring local Afghan National Police and the highway patrol, teaching them how to establish checkpoints and protect travelers. Although security along some roads improved, the Afghan population remained wary of displaying any public sign of support for the coalition and the Afghan National Army.

On 15 March 2005, the Army commenced its sixth major rotation of forces in Afghanistan. Southern European Task Force from U.S. Army, Europe, commanded by Maj. Gen. Jason K. Kamiya, began replacing the 25th Infantry Division as CJTF-76. The new command was composed of two combined task forces. The 173d Airborne Brigade from Vicenza, Italy, commanded by Col. Kevin C. Owens, became Combined Task Force BAYONET and took over RC South. It included 2d Battalion, 503d Infantry; 3d Battalion, 319th Field Artillery; 74th Infantry Detachment (Long Range Surveillance); and the 173d Support Battalion. The other task force, Combined Task Force DEVIL, took over responsibility for RC East. Commanded by Col. Patrick J. Donahue II, Combined Task Force DEVIL formed around 1st Brigade, 82d Airborne Division, and included 1st Battalion, 325th Infantry; 2d Battalion, 504th Infantry; 1st Battalion, 508th Infantry, 173d Airborne Brigade; and the 3d Battalion, 141st Infantry, Texas Army National Guard (*Chart 5*). Several Marine battalions also rotated through, and a special operations group and aviation units provided additional support. Responsibility for a fourth regional command, RC North, which had been created in October 2004 under ISAF and the Germans, changed hands in May 2005 when Italian forces took over.

Although already organized to operate as a joint headquarters, General Kamiya's command, composed of two hundred Army personnel, had to expand to over seven hundred to meet its new commitment. That increase, with about 30 percent drawn from the other American military services and the coalition, as well as individual augmentees joining in theater, posed the same continuity issues that had dogged earlier commands in Afghanistan. General Barno lamented that General Kamiya "had the disadvantage of coming in with an organization that was not, for the first time in Afghanistan, a divisional headquarters, tasked to perform a divisional headquarters and CJTF role."

What made it possible to rely on a marginally suited force such as Southern European Task Force were signs of improvement at the beginning of 2005. The focus on developing the economy and the government continued under this task force. As Lt. Col. Timothy P. McGuire, commander of 1st Battalion, 508th Infantry, 173d Airborne Brigade, explained, "The enemy can only destroy, cannot build. So our asymmetrical advantage was that we were able to bring about signs of progress." U.S. forces supported other reconstruction projects, trained the local police, and helped extend the reach of the central government. CJTF-76's medical task force sustained work started by previous rotations through Afghanistan, coordinating closely with the Afghan Minister for Public Health to assist hospitals in eastern and southern Afghanistan. Jordan, Egypt, and Spain continued to operate their hospitals in Afghanistan, and the South Koreans ran an outpatient clinic.

Chart 5—Combined Forces Command–Afghanistan: Second Rotation, Beginning April 2005

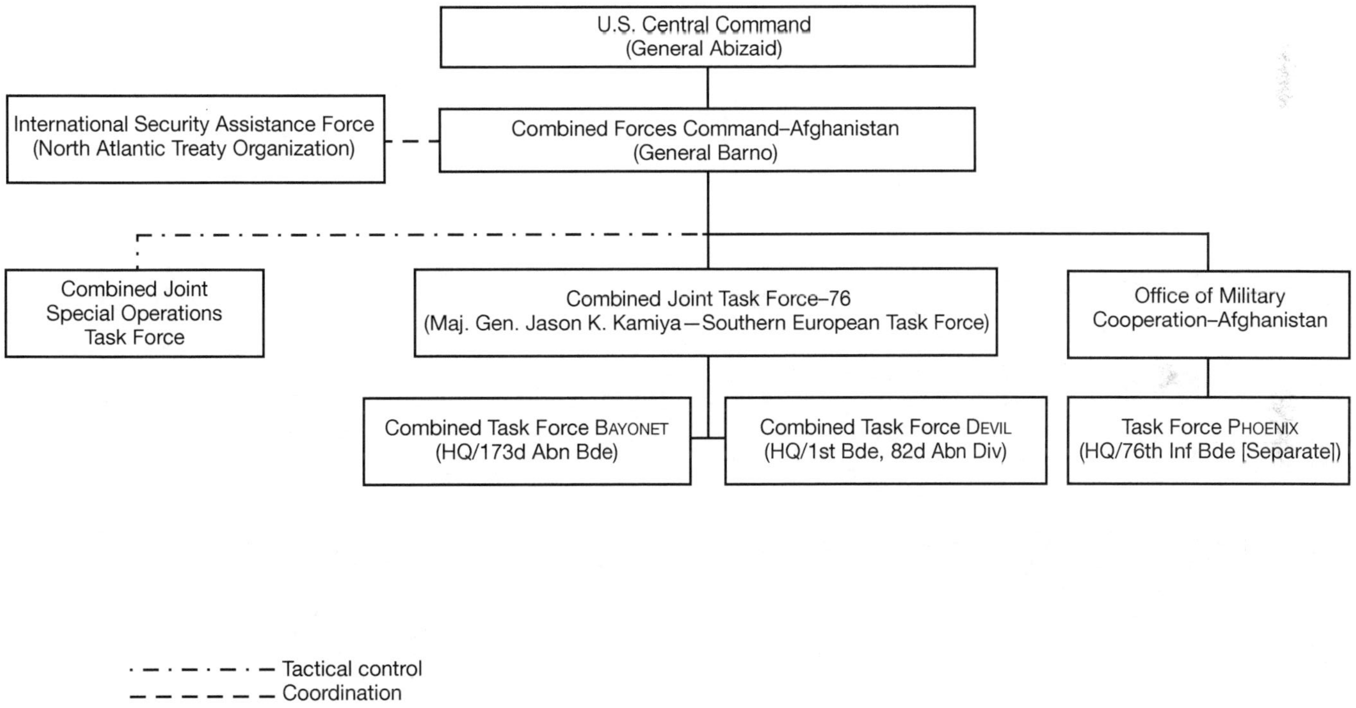

Despite continuing problems developing Afghan security forces, providing adequate humanitarian relief, and maintaining mission continuity as personnel rotated in and out of theater, the situation in Afghanistan appeared more positive at the beginning of 2005. The Afghan National Army made an increasing number of forays onto the battlefield, successful elections signified progress in forming democratic processes in a country that had little experience with them, and expanding aid to the civilian sector improved the lives of many Afghans. As General Champoux stated regarding the end of the 25th Infantry Division's tour, "The Taliban's ability to conduct operations was nonexistent." The Taliban's inability to defeat so much as a platoon, coupled with the presidential elections and the scheduled parliamentary election, "was huge."

Even with these signs of improvement, enemy attacks began to escalate in 2005. The insurgent forces made greater use of roadside and suicide bombing tactics, particularly against aid workers, Afghan soldiers, police, progovernment clerics and politicians, and those who taught classes for girls. The Taliban also started a radio station, called the Voice of Shariat. After three years of war, the Taliban remained a formidable foe and had reconstituted some of its leadership and resources. From their sanctuaries in Pakistan, Taliban leaders prepared to launch a new wave of violence on Afghanistan. With increased aid from outside sources, and often coordinating attacks with Hizb-e-Islami Gulbuddin and other anticoalition groups, the Taliban was far from defeated. The war in Afghanistan was about to enter a new, more violent phase.

Analysis

After overthrowing the Taliban regime and crippling al-Qaeda by March 2002, the United States had no intention of undertaking a long-term mission in Afghanistan. However, the need to establish a functioning state capable of preventing the Taliban's resurgence required the presence of American forces. Between 2002 and 2005, the United States, its allies, and its Afghan partners accomplished a great deal. The United States provided leadership to an international effort to rebuild a war-torn country. American forces supported a new representative government in a region that had almost no notion of the concept. They prevented insurgents from interfering with an election deemed fair and honest by international observers. They established and nurtured an Afghan army that grew in size and achieved some level of effectiveness. They adapted to an unanticipated situation and developed new doctrine and entities to fit the facts they faced on the ground. And while forced to do a great deal with very little, they achieved the objective of keeping a resilient foe at bay in an economy of force operation, while their compatriots in Iraq fought an even larger and more deadly conflict. The soldiers, sailors, airmen, and marines who participated in Operation Enduring Freedom achieved a great deal in some of the most trying conditions ever faced by U.S. armed forces.

But as evidenced by the upsurge in violence following the 2004 elections, the troubles the United States faced in Afghanistan proved far more endemic and lasting. After the invasion of Iraq in 2003, the Bush administration's reluctance to commit additional resources in Afghanistan proved to be a key impediment to American efforts. Never intending to engage in nation building and fearing involvement in the type of drawn-out struggle that had crippled the Soviet Union, U.S. policy makers sought only to create the circumstances that would allow for honorable withdrawal. This ends-based approach, which lacked a clear plan for achieving those goals, resulted in an ever-changing and often ad hoc command structure that hobbled decision making and execution in all other facets of the conflict. The failure to settle on whether to have a corps-level command separating the strategic from the tactical set the tone for pervasive uncertainty for soldiers in Afghanistan. Requiring service-oriented headquarters to become combined and joint commands through augmentation, and replacing them at intervals, forced senior leaders and their staffs to undergo a significant learning curve during their deployment. The individual services' differing rotation policies compounded the problem because they resulted in constant personnel turnover even in the midst of a command's tour in country. Some headquarters, such as the Office of Military Cooperation–Afghanistan, were created from scratch and remained understaffed for long periods. Cohesion, continuity, efficiency, and interaction with both coalition and Afghan counterparts suffered.

The approach to building a new Afghan army proved equally frustrating, with responsibility for tasks parceled out to different countries without creating a mechanism to coordinate those efforts. While there were logical reasons behind this decision, the constraints created short- and long-term deficiencies within the Afghan military. Pay remained too low to retain recruits and to minimize the likelihood of corruption, especially when the enemy offered a higher wage. Training was both too fast and too slow. Creating a battalion out of raw recruits in ten weeks produced a force unready for the test of battle. Conversely, training a single battalion at a time delayed when Afghans could shoulder more responsibility for their own security. Keeping the army small and organizing it as light infantry with limited firepower made the prospects less likely that it could deal with the enemy successfully on its own.

General Barno made an important decision in the fall of 2003 to shift the focus of operations, moving from what he and others termed *counterterrorism* to *counterinsurgency*. In the absence of service or joint doctrine to guide that campaign, he and his staff and subordinate commanders had to improvise and create their own methods. The shift in mind-set and the switch to geographic areas of responsibility were moves that, it was hoped, would better position U.S. and coalition forces for the threat posed by the Taliban. On the other hand, lacking an adequate number of troops, Barno concentrated his resources within the provinces bordering Pakistan rather than throughout the entire country. The decision to cede a sizable portion of the rural population and terrain to the enemy provided the Taliban with footholds inside Afghanistan from which it could later expand.

Regional politics also undercut coalition efforts. Pakistan's unwillingness or inability to deny extremists sanctuary within its borders prevented the coalition from destroying the Taliban. This, along with the tribal factionalism and undeveloped infrastructure in Afghanistan, gave anticoalition forces time to reconstitute and rebuild. Safe in their bases in Pakistan and hidden among almost inaccessible Afghanistan valleys, they renewed their fight against the Karzai government and its foreign supporters.

Sustaining a coalition effort in Afghanistan amid European nations' opposition to the invasion of Iraq proved troublesome. European opinion of the United States declined significantly from 2002 to 2005, with favorable opinion ratings falling below 40 percent in France, Germany, Spain, and Russia. These coalition partners had to navigate a difficult course in contributing support to American efforts in Afghanistan while dealing with public opposition to the Iraq war. Too often coalition members had to provide support in Afghanistan with caveats that secured domestic approval, which limited their ability to function as U.S. leaders desired. Differing rules of engagement and requirements for approval from home governments made employing coalition forces efficiently impossible for a theater command. Each national contingent helped, but not always as much as it might have. The need for detailed coordination also added a burden to a command and control system challenged by other shortcomings.

The United States generally, and the U.S. military in particular, had good reason to avoid getting involved in long-term nation building. But having toppled the regime in Afghanistan and wanting to prevent al-Qaeda and the Taliban from returning to power, the United States had little choice but to follow such a course in a country crippled by decades of war. The one bright spot came with the creation of the provincial reconstruction teams, an innovation in organization and process that proved effective in a limited number of areas. But more of them were needed, with more robust interagency staffing and easier access to funding.

The conflict in Afghanistan proved far more complex than envisioned in the immediate aftermath of the 11 September 2001 attacks. An apparently battered and broken enemy, low coalition casualties, and visible progress by the new Afghan government gave rise to a positive appraisal of the situation in 2005. But Afghanistan was about to require a renewed American commitment as the Taliban and al-Qaeda increased their efforts to wrest control of the country from the coalition. Operation Enduring Freedom in Afghanistan would demand a far larger and longer U.S. commitment than anticipated as victory remained an elusive goal.

Further Readings

Jones, Seth G. *In the Graveyard of Empires: America's War in Afghanistan*. New York: W. W. Norton, 2009.

Rashid, Ahmed. *Taliban: Militant Islam, Oil and Fundamentalism in Central Asia*. New Haven, Conn.: Yale University Press, 2000. Second edition published in 2010.

Wright, Donald P., et al. *A Different Kind of War: The United States in Operation ENDURING FREEDOM (OEF), October 2001–September 2005*. Fort Leavenworth, Kans.: Combat Studies Institute Press, 2010.

Abbreviations

CENTCOM	U.S. Central Command
CFC-A	Combined Forces Command–Afghanistan
CFLCC	Combined Forces Land Component Command
CIA	Central Intelligence Agency
CJSOTF	Combined Joint Special Operations Task Force
CJTF	Combined Joint Task Force
CTF	Combined Task Force
FBI	Federal Bureau of Investigation
FM	field manual
HHC	Headquarters and Headquarters Company
ISAF	International Security Assistance Force
NATO	North Atlantic Treaty Organization
NGO	nongovernmental organization
NOFORN	not releasable to foreign nationals
PRT	provincial reconstruction team
RC	Regional Command

All photos, except for the cover, are in the U.S. Army Center of Military History files.

CMH Pub 70–122–1

Cover: *Soldiers from Company B, 1st Battalion, 505th Infantry, conduct a dismounted patrol in the mountains of Afghanistan on 7 October 2004.* (Department of Defense)

Made in the USA
Monee, IL
08 November 2026

67380791R00044